RAND NATIONAL DEFENSE RESEARCH INSTITUTE

T0108948

Technological Lessons from the Fukushima Dai-Ichi Accident

Cynthia Dion-Schwarz, Sarah E. Evans, Edward Geist,
Scott Warren Harold, V. Ray Koym, Scott Savitz, Lloyd Thrall

Prepared for the Office of the Secretary of Defense
Approved for public release; distribution unlimited

For more information on this publication, visit www.rand.org/t/RR857

Library of Congress Cataloging-in-Publication Data is available for this publication.
ISBN: 978-0-8330-8827-7

Published by the RAND Corporation, Santa Monica, Calif.
© Copyright 2016 RAND Corporation
RAND® is a registered trademark.

Support RAND
Make a tax-deductible charitable contribution at
www.rand.org/giving/contribute

www.rand.org

Preface

In March 2011, northern Japan was subjected to a devastating earthquake and tsunami. One of the many secondary effects of these disasters was a loss of control of the Fukushima Dai-Ichi nuclear plant. This led to the dispersal of radioactive material into the environment, both immediately and over the following months. The focus of this research is on how various technologies were used to ascertain the extent of radioactive contamination, to prevent the spread of radioactivity that had already dispersed into the wider environment, to decontaminate areas or items, and to store radioactive material for extended periods, all while limiting human exposure to radiation. By capturing lessons regarding how technologies were used successfully, as well as identifying capability gaps that could have been alleviated through the better use of technologies or the development of novel technologies, this research is intended to help improve technological preparedness for any future radiological or nuclear incidents.

This research was sponsored by the Office of the Secretary of Defense and conducted within the Acquisition and Technology Policy Center of the RAND National Defense Research Institute, a federally funded research and development center sponsored by the Office of the Secretary of Defense, the Joint Staff, the Unified Combatant Commands, the Navy, the Marine Corps, the defense agencies, and the defense Intelligence Community under Contract W91WAW-12-C-0030.

For more information on the RAND Acquisition and Technology Policy Center, see www.rand.org/nsrd/ndri/centers/atp or contact the director (contact information is provided on the web page).

Contents

Summary

Following the earthquake and tsunami in Japan on March 11, 2011, through December 2013, RAND analyzed how technologies had been used to address radioactive environmental contamination caused by the loss of control of the Fukushima nuclear power plant. The goal of this study is to analyze lessons from the Fukushima experience so that the U.S. Department of Defense (DoD) can be better prepared to respond to nuclear or radiological emergencies in the United States or overseas.[1] Given prospective DoD needs and responsibilities following an incident, DoD stakeholders were keenly interested in understanding how technologies contributed to all of the following at Fukushima:

- characterizing the extent of contamination over space and time
- prevention and limitation of human exposure

---

[1] For example, in the event of an attack in the United States using a nuclear weapon or a radiological dispersion device, DoD would likely be called upon to provide defense support to civil authorities (DSCA). If the attack happened in the vicinity of a military base, DoD would also need to respond to protect its own personnel and assets. In addition, DoD could be called upon to provide DSCA following a major release from a nuclear power plant in the United States, an accident involving nuclear weapons, or a leak from a nuclear-powered aircraft carrier or submarine. Overseas, DoD could be called upon to help respond to nuclear or radiological attacks against allied nations or U.S. bases within them. It may also need to respond to releases of radioactivity on battlefronts or in captured areas. Finally, DoD may help to respond to an accidental release overseas, as it was following the devastating earthquake, tsunami, and nuclear accident in Japan (DoD's response was termed Operation Tomodachi, meaning "friend" in Japanese). See Federal Emergency Management Agency (FEMA), "Nuclear/Radiological Incident Annex," June 2008; and FEMA, "Federal Radiological Preparedness Coordinating Committee," FEMA website, updated June 26, 2013. Also see U.S. Northern Command, "Joint Task Force Civil Support," website, undated.

- decontamination operations
- disposal of radioactive material.

DoD was especially focused on understanding what gaps in response capabilities could have been addressed using technologies that are either not yet available or that have not yet been integrated in a way that would make them useful in the context of a radiation-dispersal incident. Potential investment in such technologies can make DoD more capable in responding to future radiological or nuclear incidents.

Key Themes and Findings

We found three recurring themes throughout our research:

1. **Response capability requirements will be diverse.** Specific response capability requirements will vary considerably from one incident to another, particularly if the incident is of a different type (for example, a nuclear detonation, rather than loss of control at a nuclear plant). Requirements can also be highly localized over space and time within a single incident.
2. **The scale of response required is a major driver behind the challenges involved.** Given the vast areas and quantities of material affected by a large-scale incident, cost-benefit analyses are imperative.
3. **Public perceptions will be a major factor shaping response.** Even the most scientifically sound approaches cannot be implemented in the face of public opposition due to popular misperceptions or a loss of trust.

As a result of our examination of the technological response to the Fukushima disaster, we found several promising areas for further technological development, along with a cautionary note regarding the use of some technologies:

- Distributed, wide-area sensors for rapid, real-time measurement—along with the employment of information technology to share data—will be critical for an effective response.
- Unmanned systems—especially ground systems that are able to negotiate obstacles and operate autonomously in hazardous areas—are needed in austere, contaminated environments.
- Personalized protection—hazmat suits, personal dosimetry, and medicine—will enable fuller and safer responses after a nuclear disaster.
- Cost-effective technologies to decontaminate and store large quantities of water, land, and artificial surfaces would most certainly be a game-changing set of technologies in such an event.
- Public perceptions will powerfully shape the response to such a disaster—and may preclude some feasible and affordable technological solutions.

The Events Leading to the Contamination

When the massive 9.0 moment magnitude[2] earthquake shocked the Tohoku region on March 11, 2011, engineers and technicians immediately initiated emergency shutdown procedures on the operating reactors at Fukushima Dai-Ichi nuclear power station, about 160 kilometers away from the earthquake's epicenter. The reactors, although suffering some damage, were built to withstand the large earthquakes common in Japan, with emergency backup power generators supplying electricity to operate crucial water-circulation equipment used to cool the fuel. Boron rods were automatically inserted into the fuel to stop the chain reaction, and operators initiated emergency water circulation powered by backup diesel generators. Then, an hour after the earthquake, tsunami waves up to 15 meters in height breached the 6-meter sea wall at Fukushima, and all power was lost, including the backup systems.

---

[2] Readers may be more familiar with the Richter scale as a measure of an earthquake's magnitude. Professional geologists have used the *moment magnitude* measure since about 1980 as a more precise measure of an earthquake's energy release. The scales are quite similar, so a 9.0 moment magnitude earthquake is indeed a very large, once-in-a-generation event.

With the loss of power after the tsunami, passive heat generation by the nuclear fuel continued, and the cooling water in reactors began to super-heat and corrode the fuel, thereby generating hydrogen gas. By day four, three of the reactors had experienced meltdowns and hydrogen explosions in the buildings housing several of the reactors, resulting in contamination of the surrounding countryside. Nuclear fission products—primarily comprising radioactive tellurium (half-life 14 seconds), iodine (half-life 8 days), and cesium (half-life 2 years or 30 years, depending on the isotope)—spread by the plants' explosions contaminated the surrounding farmland, trees, villages, and ground and sea water. The tellurium and iodine rapidly decayed into harmless byproducts, but the cesium persists.

The Japanese government is committed to cleaning up the immense area and will fully decommission the plant at Fukushima Dai-Ichi over the next several decades. This report discusses the successes and failures of the technologies used in the cleanup.

Characterizing the Extent of Contamination and Personal Exposure

Knowing the extent and nature of contamination as a function of time and space is essential for minimizing human exposure, designating the extent to which areas or resources can be used, and undertaking effective decontamination efforts. While modeling can contribute greatly to such understanding, it needs to be supported and updated by data collection, given uncertainty about the quantity of radioactive materials released and their patterns of migration through the environment. Effective characterization can reduce the risk to both responders and the general populace and, in doing so, may boost public confidence in the response effort. It can also reduce the amount of material, time, and manual labor required to mount an effective response, as it reduces duplication of effort and increases efficiency.

Shortly after the Fukushima event, a number of aerial surveys were conducted to rapidly enable broad decisionmaking regarding response. However, these aerial surveys had limited granularity in characterizing local variations in radiation emissions, so subsequent ground surveys were necessary to provide more-detailed characterization. Radiation levels can vary appreciably over distances of as little as a few meters, due to topography, biological concentration, and other environmental factors. Radiation levels also change dynamically over time, as radioactive particles both migrate and decay. Further, air conditions (particularly wind, precipitation, and particulate matter) can interfere with reliable and granular airborne characterization. Finally, an aerial survey can misinterpret higher-altitude areas, which are closer to the aircraft, as being more highly contaminated than lower-lying areas with the same or greater degree of contamination. Moreover, aerial surveys are resource-intensive and costly; experts whom we interviewed cited a desire for a larger number of simpler, less-expensive methods.

Unfortunately, ground surveys by vehicles and robots were greatly impeded by the destruction of the area's infrastructure, including fuel supplies, road networks, electricity, potable water supplies, and communication networks. Despite these problems, both personnel in vehicles and robots conducted a great deal of ground survey work. Unfortunately, the data sets collected by these surveys were not integrated into a common operational picture. Overall, gaps in knowledge of radioactive contamination contributed to both response shortfalls and additional risk to personnel.

In the event of a future attack or accident releasing radiation, DoD would likely be called upon to help characterize radiation levels (as it did in the aftermath of the Fukushima incident). In many possible circumstances, the affected area's physical infrastructure would be degraded due to a concomitant disaster or attack, as it was in Fukushima. As such, aerial surveys would likely be a useful initial source of information. Moreover, based on the Fukushima experience, there is a need for rapidly deployable distributed sensors that can report continuously on local radioactivity and wind conditions. Such sensors could be situated on inhabited or unmanned vehicles, or distributed from such vehicles onto the ground and other surfaces. Naturally, the sensors would need to be small, rugged, low-cost, radiation-hardened, and long-endurance. They would need to provide low-bandwidth transmissions to relay platforms, such as unmanned aerial vehicles or aerostats, so that they can provide actionable data in real time to analysts and modelers. Personal dosimeters would also be a useful part of this network, ideally providing both real-time information to those wearing them, as well as situational awareness for decisionmakers. To that end, Japan's Nuclear Regulation Authority is developing an advanced dosimeter that can assess exposure on an hourly basis, allowing individuals to better protect themselves in real time and relaying data to decisionmakers to identify higher-radiation areas. DoD use of similar networked dosimeters could help to improve commanders' situational awareness and reduce personnel exposure. Ideally, such dosimeters would also be able to provide radiation readings at shorter intervals, so that the personnel wearing them could respond accordingly.

Robotics

Ground robots were used extensively to survey the environment at Fukushima, as well as to try to regain control of the plant and conduct decontamination in select areas. However, robots were frequently impaired by all of the following issues, particularly in or near the nuclear plant:

- Mobility limitations arose when robots were faced with obstructions such as rubble piles, tree roots, stairs, and doors, as well as the robots' tangled cords.
- There were anecdotal reports that a lack of radiation hardening resulted in degraded electronics and cameras within a few meters of highly radioactive material.
- Mission durations were short due to the limited battery lives of untethered robots, as well as radiation damage.
- Blocked transmissions caused by buildings and degradation of communication networks created difficulties in communicating with robots. In the absence of a high degree of autonomy, this constrained their ability to act or move, as well as to relay information. In many cases, data streams could not be collected from a

ground robot until it had been physically recovered; further time was then required for data analysis.

- Within the most difficult areas of the facility, robots developed an inability to function under exceptionally harsh environments. In addition to high radiation, robots were subjected to darkness, severe heat, unpredictable surroundings, and channelized movement. Furthermore, they had to be able to operate in both open air and underwater environments, often having to conduct nimble movements simulating those of a human hand.

In cases where ground mobility is needed, advances in robot autonomy, the ability to overcome or bypass obstacles, and having dexterity and endurance would be invaluable. Greater autonomy and data-analysis capabilities could reduce either the bandwidth associated with communications or the frequency with which the robot needed to communicate. Moreover, these advances would be useful in a variety of civilian and military contexts. For this specific type of disaster response, developing radiation-hardened robots also may be needed; this could be done through shielding of their electronics and redundancy in circuit design.

Radiation Suits and Collective Protection

Personnel who need to operate in contaminated environments obviously require protection appropriate to the amount and types of radiation in their vicinity, as well as the duration of their exposure. At Fukushima, sources noted that the weight and design of existing suits caused rapid heat fatigue, nausea, and intense headaches, making personnel less capable and limiting the amount of time they could wear the suits. These would also be issues for DoD personnel responding to a radioactive or nuclear incident.

Potential ways of addressing this issue include the following:

- the use of exoskeletons—suits that are more like external armor than clothing—potentially designed to enhance human strength,

rather than sap it. These could also offer protection against external exposure, whereas most radiation suits are currently designed
to prevent inhalation of particles.
- if possible, the development of lighter-weight materials that could
perform the same function as the existing suits.
- the use of collective-protection shelters as a substitute for individual suits. These could be placed in well-surveyed and decontaminated "islands," perhaps connected by similarly well-surveyed
"corridors," as bases for operations. Alternatively, self-contained
biospheres that provide collective protection could be placed even
in areas that had not been decontaminated. Conversely, encap-
sulating "hot spots" that are irradiating a wider area by putting
deployable barriers around them can diminish the need for individual suits.

Medical and Genetic Aspects of Radiation Health Effects

Health physicists disagree about the degree to which some individuals
may be more or less sensitive to the effects of radiation than others, and
more research is needed to answer this question definitively. If such
differences exist, they might enable screening techniques to prevent
individuals with unusually high radio-sensitivity from living or working in radiologically compromised areas. Furthermore, insights from
such research could form the basis for either gene therapies or new
radio-protective drugs to reduce individuals' vulnerability to the health
effects of radiation, reducing the risk to military personnel who may
encounter hostile radiation environments. Obviously, costs and acceptance would be critical issues. Likewise, regenerative medicine provides another potentially promising area of research to ameliorate the
health effects of radiation exposure. Interview subjects suggested that
stem-cell research might prove a fruitful avenue of research to this end,
proposing, for instance, that the availability of regenerative medicine
could increase the willingness of individuals to work in contaminated
areas. DoD could pursue advances in these areas to improve effectiveness and reduce vulnerability in a radioactive environment. Naturally,

the perceptions of military personnel and the general public would be important considerations in this context.

Minimizing Exposure via Food

Shortly after the accident, ingestion of radioactive material became the primary human health risk. The Japanese developed a range of technologies to combat these hazards. Every bag of rice grown in the surrounding area was tested for radiation using an assembly-line system, and only a few bags were found to emit more radioactivity than was conservatively deemed acceptable for human consumption. Fish from the area are now being assessed using a non-destructive testing regimen developed by Tohoku University. Monitoring 100 percent of foodstuffs would restore greater confidence and enable the local economy to recover more robustly, but challenges of cost and scale make this goal difficult. Furthermore, the local agricultural economy may suffer from stigmatism even after robust food testing, as both local and export markets increase consumption from areas never exposed to contamination. While technology can help containment, cleanup, and verification, a full recovery of the local agricultural economy may prove a difficult and costly process, one that depends on public perceptions as well as technical achievements.

In the aftermath of a radiological or nuclear incident, DoD would likely face concerns about the safety of food for its personnel and the populations it was supporting. One easy way of addressing this problem would be to bring in food, such as ready-to-eat meals, also known as MREs, from outside the area. Over the long term, as part of an effort to rehabilitate the area economically, DoD personnel and those whom they are supporting would likely consume some local foods. In that context, a few simple approaches can help to reduce exposure. These include using especially potassium-rich fertilizers (which help to prevent plants from taking up cesium), feeding livestock absorptive materials that will help to keep radionuclides from being absorbed into their bodies, and avoiding certain foods that are cesium-accumulators, such as mushrooms. The assembly-line regimens developed for Fukushima can also be applied to test food before it is eaten.

Decontamination

One of the most challenging aspects of response to an incident like that at Fukushima is decontaminating the environment to a sufficient extent that human activities can resume. At Fukushima, widespread cesium contamination persists in the soil, plants, and urban areas surrounding the plant; at the plant itself, water is also contaminated. This variety of contaminated materials complicates the decontamination approach. However, broadly speaking, there are three approaches to decontamination:

- physical decontamination—removing surface-borne or airborne radionuclides by applying direct mechanical force and/or using flowing water or air. At Fukushima, this was done using relatively simple tools, such as shovels, brushes, rags, hoses, street-sweepers, vacuum filters, and concrete shavers. The vast majority of decontamination in the region around Fukushima was performed using these labor-intensive physical methods, accentuating worker-health issues relative to public-health issues.
- chemical decontamination—taking advantage of atomic-level interactions to concentrate radionuclides of particular elements into a smaller mass of material. Near Fukushima and elsewhere, solid materials have been chemically decontaminated using liquids, foams, or gels to extract radionuclides. Water can have dissolved radionuclides removed through the use of molecular sieves, or by adding materials that cause the radionuclides to precipitate out of solution.
- biological decontamination—employing living systems' preferences for absorbing the atoms of particular elements to concentrate radionuclides from a large medium (such as soil or water) into a smaller mass of organic material (such as sunflowers, algae,

cannabis, or mushrooms). This has been done near Fukushima on a limited scale, using sunflowers.

Since radioactive material needs to be secured for periods ranging from decades to millennia, minimizing the mass of waste involved is imperative. In general, physical decontamination processes are the least selective in concentrating waste and generate the largest quantities of contaminated material, while biological ones are the most selective, generating smaller quantities of more highly radioactive material.

The area surrounding the Fukushima Dai-Ichi plant was an especially challenging environment in which to conduct decontamination: forest and mountain environments retain radioactive materials much better than do most urban environments due to cesium capture attributes of the natural clay in soil. However, even in less challenging environments, there is currently no economically efficient way to decontaminate large areas or large amounts of human infrastructure. Only small, prioritized areas can be decontaminated, and even in those areas, there will almost always be some residual contamination. Moreover, even if residual contamination has a near-zero impact on health, public concern will likely limit future use of the area or require costly programs to further reduce already marginal radiation levels.

Moreover, since the nuclear plant needed to be continually cooled, vast quantities of contaminated water was (and is) created, particularly at the plant. Certain radioisotopes (such as cesium-137 or strontium-90) that have been dissolved in water can be precipitated out by adding materials to the water, enabling the precipitated particles to be removed via filtration. However, this does not address the problem of water molecules in which one or both of the hydrogen atoms is tritium, a radioactive form of hydrogen. Since radioactive tritium water molecules have essentially the same chemical properties as other water molecules, separating them is highly energy-intensive, and cost-prohibitive on a large scale.

DoD's role in long-term decontamination efforts would likely be limited, unless such efforts were being undertaken on a base. However, should DoD anticipate the need to conduct decontamination, there are several approaches for consideration. One would be to develop plants, algae, or even bacteria that would have a greater preference for cesium than existing organisms do, to help in removing radioactive cesium from the soil or water. Ideally, these organisms could also be designed to incorporate other elements whose radionuclides might also be distributed by an event, such as strontium and uranium. The organisms can then be collected and desiccated, generating much smaller quantities of more concentrated waste than would be involved in removing the topsoil. For man-made surfaces, a complementary approach would be the development of robots for surface decontamination; our contacts noted that a number of robots are currently being developed or tested for this purpose. Such robots could blast high-pressure water, dry ice, or tiny balls (smaller than a millimeter in diameter) at contaminated surfaces to remove superficial contamination, and then vacuum up the liberated material for disposal. Finally, the ability to vitrify some radioactive materials (i.e., encase them in glass) *in situ*, at a reasonable cost and without damaging ecosystems, would obviate the need to remove and transport them.

Disposal

Disposing of contaminated material is one of the most challenging aspects of responding to radiological disasters. This material must be identified, collected, and then managed in a way that minimizes health and ecological risks, with all of these activities tailored to the specific incident and materials in question. In the case of Fukushima, disposal of contaminated soil, leaves, and other solid materials has been extremely difficult, due to the huge volumes of material involved and the lack of any politically acceptable means of handling it. Ideally, it would be possible to process this material to concentrate the radionuclides it contains, but this is obviously difficult to accomplish in heterogeneous solids. As such, it is unclear where ultimate storage would occur; even well-stored materials generating minimal health risks may face resistance from local communities.[3] Without adequate space in nuclear waste disposal facilities to accommodate this material, it has accumulated at sites near where it was collected, awaiting a permanent solution. In many instances, accumulating the material may have been counter-productive—it would have been better to leave it in place, allowing its radioactive contents to disperse and decay—but political imperatives required its collection, during which time personnel were exposed to its harmful effects.

Disposal of contaminated water at Fukushima has also been challenging for both technical and political reasons, as well as the immense quantities of water involved. As noted above, there is no efficient way to remove tritium from water, so the entire volume of water must be disposed of. Diluting the tritium-laden water in the vastness of the Pacific would pose a negligible health and environmental risk, but would be very unlikely to be politically acceptable due to both domestic and international public perception. Domestic and international laws could also curtail this approach.

To the extent that the U.S. government (likely not DoD) would need to dispose of radioactive material after a possible future incident, it would most likely find a geologically stable repository on land. Alternative approaches that have been suggested include depositing such waste in the ocean depths or even in outer space, though the latter would likely be both prohibitively expensive and risky. Another alternative that has been proposed, transmutation of elements, would be prohibitively energy-intensive and expensive if done on any meaningful scale. Regardless of how material is disposed of, the importance of devising solutions that will be acceptable to the general public cannot be overstated.

---

3   The idea of storing nuclear waste in a storage facility at Yucca Mountain, in a sparsely populated area of the United States, was so controversial that it has been terminated, although it could be revived at some future date. The political difficulties of developing a similar storage site in densely populated, earthquake-prone Japan, particularly after the Fukushima incident, would likely be far greater.

Conclusions

DoD may be required to address a radiological or nuclear event some point in the future, as it was in the case of Fukushima. To successfully confront these challenges on a mass scale, response measures must be cost-effective. While promising emerging possibilities exist for radiological surveillance, decontamination, and disposal, considerable research and development is necessary to make these technologies ready for military deployment. Current decontamination and disposal techniques are relatively ineffective and expensive, which severely limits the ability of states to ameliorate the consequences of radiological incidents. Furthermore, both political and popular acceptance are essential for any measures envisioned for DoD use, and development decisions should take this consideration into account.

Acknowledgments

Japan's response to the March 11 "Triple Disaster" drew upon the efforts of exceptional human beings: engineers forging solutions under incredible pressure, doctors driving into an undetermined radiological disaster to administer treatment, and first responders risking their lives to assist and care for strangers, to name a few. One cannot do first-hand research into this subject without being struck by the selflessness, dedication, and competence of such individuals, who fit any definition of the word "hero." We are grateful for their willingness to share their experiences and generate the lessons contained within this report.

This research would not have been possible without the many academics, scientists, engineers, first responders, military officers, and policymakers who volunteered their time and expertise to discuss this sensitive subject. On both sides of the Pacific, the candor and competence of these professionals shed new light on areas for potential research and development in the fields of nuclear security and radiological preparedness. As such, they have contributed to the safety and security of citizens broadly, helping to ensure the lessons learned in tragedy are transferred to the institutions, communities, and experts who can transform them into better solutions. While we hope the need to apply such lessons does not arise, the RAND team felt privileged to assist in turning their firsthand experience of disaster into tangible lessons for the public good.

The conditions of anonymity necessary to candidly discuss such sensitive issues preclude publishing most of the organizations and experts consulted in creating this report. However, we would like to express our gratitude to the members of the Diet, the Japanese Government, Tokyo University, Mitsubishi Research Institute, and U.S. Embassy Japan who gave generously of their time and insight. We would also like to thank Col Edward Vaughan, U.S. Air Force, for his insights into Operation Tomodachi and bilateral cooperation. Finally, we would like to express our particular gratitude to the administrators and engineers at Tokyo Electric Power Company, who gave us first-hand insight into the immediate and ongoing recovery efforts.

Finally, we are grateful to Richard Danzig, who provided extensive comments on the approach, and to Dr. Robert Sindelar of the Savannah River National Laboratory and Dr. Don Snyder of the RAND Corporation, who gave thoughtful reviews to the final report.

In the course of researching this report, RAND spoke with many experts who expressed their frank, and personal, views to us. We thank them, and do not mistake their views as representative of their own organizations. In addition, RAND interviewed these individuals after undergoing a full Human Subject Protection review of our protocols, in compliance with all applicable federal statutes and DoD policies governing Human Subject Protection.

Abbreviations

DoD	U.S. Department of Defense
DSCA	defense support to civil authorities
FEMA	Federal Emergency Management Agency
GPS	Global Positioning System
HEPA	high-efficiency particulate air
JSDF	Japan Self-Defense Forces
NRA	Nuclear Regulation Authority (Japan)
TEPCO	Tokyo Electric Power Company

CHAPTER ONE

Introduction

The Events Leading to the Contamination

When the massive 9.0 moment magnitude earthquake shocked the Tohoku region on March 11, 2011, engineers and technicians immediately initiated emergency shutdown procedures on the operating reactors at Fukushima Dai-Ichi nuclear power station, about 160 kilometers away from the earthquake's epicenter. The reactors, although suffering some damage, were built to withstand the large earthquakes common in Japan, with emergency backup power generators supplying electricity to operate the crucial water-circulation equipment used to cool the fuel. Automated control systems inserted boron rods into the fuel to stop the chain reaction, and initiated emergency water circulation powered by backup diesel generators. Then, an hour after the earthquake, tsunami waves up to 15 meters in height breached the 6-meter sea wall at Fukushima, and all power was lost, including the backup systems.

Fukushima Dai-Ichi's six boiling-water reactors use uranium fuel pellets (half-life 700 million years) encased in zirconium, with long boron control rods to absorb the neutrons and stop the controlled chain reaction. It takes moments to automatically insert the rods into the fuel, but the radioactive fission products continue to produce heat. Complete shutdown entails insertion of the rods (completed within minutes of the earthquake) and removal of the fuel and products to cooling ponds (not initiated since this can take days); it can take weeks to fully shut the reactors down. In its intermediate shutdown state, water is pumped through the system to extract heat and, crucially, prevent meltdown. Prior to the earthquake, reactors 1, 2, and 3 were fully operational, supplying power to eastern Japan, with reactors 4, 5, and 6 partially or fully shut down for normal maintenance. With the loss of power after the tsunami, passive heat generation by the nuclear fuel continued, and the cooling water in three of the six reactors began to super-heat. At high enough temperatures, water reacts with zirconium resulting in volatile hydrogen gas.

Tokyo Electric Power Company (TEPCO), the plant owner, improvised a number of solutions in their struggle to cool the fuel and waste and vent the dangerous hydrogen gas over the next few days. The earthquake had destroyed roads and the tsunami had wiped out much of the local equipment, so resources were extremely limited. By day four, three of the reactors had experienced meltdowns and hydrogen explosions, spreading radioactive nuclear fission products[1] over an area the size of Connecticut. At the plant itself, radiation levels began rising and TEPCO evacuated all but a few select volunteers, who became known as the Fukushima 50.

Meanwhile, the Japan Self-Defense Forces (JSDF), local responders, and the international community stepped in. Volunteers from a JSDF helicopter wing dropped water onto the reactors from above; firefighters gave themselves an hour on site to rig hoses; and Navy ships sprayed seawater onto the plant. The Japanese government was concerned that events could cascade out of control and result in a vastly larger radiological release that might threaten Tokyo, 260 kilometers away. But by the end of March, the responders had successfully prevented a catastrophic escalation of the disaster and radiation levels dropped sufficiently for workers to return to the plant. Fears of further nuclear radiological explosions subsided. Workers at the plant continue to struggle daily to contain massive quantities of radioactive water, but local power has been restored and attention has widened to grapple with decommissioning the plants and mitigating the environmental disaster.

---

[1] The nuclear fission products, as opposed to the nuclear fuel, are the byproduct of the heat-generating controlled nuclear fission of the uranium fuel. This waste comprises various elements, and is often itself radioactive.

The Japanese government evacuated over 100,000 people from the area surrounding the plant. Nuclear waste—primarily comprising radioactive tellurium (half-life 14 seconds), iodine (half-life 8 days) and cesium (half-life 2 years or 30 years, depending on the isotope)—spread by the plant's explosions contaminated the surrounding farmland, trees, villages, and ground and sea water. The tellurium and iodine rapidly decayed into harmless byproducts, but the cesium persists—a nasty, sticky metal that chemically mimics potassium in biological systems. For historical reasons, the Japanese people are particularly concerned about long-term radiation exposure, so the locals now routinely scan foodstuffs from the region for cesium signatures. The market for Geiger counters in Japan grew 20-fold after the disaster.[2]

There have been no confirmed causalities as a result of acute radiation poisoning in the months following the disaster,[3] and the World Health Organization estimates that increases in cancer rates due to the contamination will be undetectable.[4] Although uranium fuel melted down at the plant, it was not spread over the environment: only the somewhat-easier-to-deal-with and somewhat-less-dangerous nuclear waste was spread. The Japanese government is committed to cleaning up the immense area and will fully decommission the plant at Fukushima Dai-Ichi over the next several decades. This report discusses the successes and failures of the technologies used in the cleanup.

Motivation

Following the earthquake and tsunami in Japan on March 11, 2011, through December 2013, RAND analyzed how technologies had been

---

[2]   Pavel Alpeyev, "Japan Geiger Counter Demand After Fukushima Earthquake Means Buyer Beware," *Bloomberg News*, July 14, 2011.

[3]   The World Health Organization estimates that whole-body doses of radiation that people near and around the site were exposed to never exceeded 50 millisieverts, well below the deadly acute dose of 10,000 millisieverts. World Health Organization, *Health Risk Assessment from the Nuclear Accident After the 2011 Great East Japan Earthquake and Tsunami*, 2013.

[4]   World Health Organization, 2013, p. 8.

used to address radioactive environmental contamination caused by the loss of control of the Fukushima nuclear power plant, especially in helping the U.S. Department of Defense (DoD) prepare to respond to similar events in the future. In particular, DoD stakeholders were keenly interested in understanding how technologies contributed to all of the following:

- characterizing the extent of contamination over space and time
- decontaminating surfaces, infrastructure, the environment, and personnel
- storing, transporting, and ultimately disposing of radioactive material.

DoD stakeholders were particularly focused on understanding what gaps in response capabilities could have been addressed using technologies that are either not yet available or that have not yet been integrated in a way that would make them useful in the context of a radiation-dispersal incident. Such events can include small-scale releases from nuclear power plants, large-scale loss of containment at nuclear power plants, radiological attacks, or nuclear attacks.[5] DoD can potentially develop such technologies in an effort to better address future radiological or nuclear incidents in the United States or overseas.

We found four broad themes that recurred throughout our research, and we would like to highlight them at the outset; we will return to them repeatedly through the course of the paper.

1. **Response capability needs will be very diverse.** All response capabilities need to address highly local conditions and needs. Different areas will require different types of response, as will different media such as soil, concrete, water, and air. Moreover, the same area will often require different types of response capabilities as a function of time. This diversity extends to incident types, as well: different response capabilities will be needed for a

---

[5] A radiological attack would involve the dispersion of radioactive material without a nuclear reaction; for example, radioactive material could be packed around conventional explosives to disperse it over a wide area.

radiological weapon, a nuclear weapon, a small-scale accidental release, and a larger, ongoing accidental release.

2. **Many of the problems associated with response to a large-scale incident of this type are a function of the considerable scale involved in response, which necessitates thorough evaluation of the benefits and cost-effectiveness of particular measures.** A few gallons of water or square meters of soil could be remediated at acceptable costs; the vast quantities and areas involved in a large-scale incident are much less susceptible to cost-effective solutions. (We are defining costs broadly, to include human health and impact on society.) Decontamination is very hard on a large scale, as is safe and secure storage of radioactive material.

3. **The widespread and timely dissemination of actionable, clear, transparent, credible information is critical to reducing effects and societal costs, as well as shaping response.** People can take better actions to protect their own health, and economic effects are likely to be diminished, if the public has been well-informed by trusted sources. In addition, public perceptions have a powerful influence over response decisions; scientifically sound approaches need to be accepted by the wider public if they are to succeed.

4. **A highly varied "toolkit" of capabilities is needed for effective response.** Given the diversity of response capabilities needed on large scales, as well as numerous constraints due to costs and public perceptions, a great variety of approaches are needed to address radiological or nuclear events: no one approach will be able to be widely applied. Fortunately, there is already a wealth of research on radiological and nuclear issues in both the civilian and military sectors. One role for DoD researchers is to think about how to leverage such research, while also trying to think about how civilian and military capabilities can complement one another.

Organization of This Report

We have structured this report around four principal aspects of response to radioactive contamination, with each of the following being the subject of a chapter:

- characterizing the extent of contamination
- preventing further damage by, or dispersion of, radioactive material
- decontamination and collection
- safely and securely disposing of contaminated materials.

Within each of these chapters, we begin by broadly discussing what this type of response entails, then focus on the experience at Fukushima. Based on problems and successes from that experience, we then highlight the prospective direction of solutions to future challenges.

Next, we explore the use of robots in Fukushima, and lessons derived from that experience. Although robots are always associated with a particular mission, including those listed above, we felt that there was such an abundance of material regarding robots that was not mission-specific that it was best to treat them separately. After this, we briefly review the experience of responding to the Chernobyl incident of 1986. This helps to highlight advances since that event, as well as similarities between the two types of response. Moreover, the quarter century since the Chernobyl incident has provided more time for lessons to be gathered about long-term response issues than are available for the Fukushima case.

Finally, we present some conclusions and recommendations based on our analysis.

CHAPTER TWO

Characterizing the Extent of Contamination

To address a radiological incident effectively, it is critical to ascertain both how much radioactive material has been released and what has been exposed—people, infrastructure, equipment, and the larger environment. This knowledge is essential for minimizing additional human exposure, designating the extent to which areas or resources can be used, and undertaking effective decontamination efforts.

There are many complex aspects of this assessment process. In general, radiation measurements will be highly localized: the amount of radiation exposure may vary considerably over relatively small areas. For example, wherever slight variations in local topography cause puddles of water to form, radiation collected by the water will be concentrated toward the center of each puddle as it evaporates, leading to "leopard-spot" patterns of radioactivity. For any given point on the Earth's surface, exposure to radioactivity will vary strongly as a function of altitude, depending on the degree to which radioactive material is confined to the surface, is intermingled with sub-surface soil or water, is distributed in the air, and/or is present on nearby surfaces that vary in height (such as trees or buildings). Radiation measurements will also change dynamically over time, as radioactive particles both migrate and decay. Wind, precipitation, and other environmental phenomena will cause radionuclides to migrate, as will human and animal activities. Some biological systems have a particular propensity to concentrate certain radionuclides,[1] and concentrations may be correlated

---

[1] Really, they concentrate specific chemical elements, such as cesium or strontium, including the radioactive isotopes of those elements that are of concern. See, for example, P. Soudek,

with organisms' positions on the food chain. In addition, a number of processes affecting radiation dispersal throughout the environment are likely to be cyclical. For example, radionuclides may be concentrated in the leaves of trees during spring and summer, only to later be redispersed in the environment when the leaves fall to the ground during autumn and winter.

In general, to track dispersal of radioactivity accurately, there are several key data sets that need to be collected at any given point in time; moreover, given the considerable changes that will occur as conditions evolve, these data sets will require both additional measurements to ensure their continued accuracy as well as modeling of their change over time as they decay or migrate. Naturally, decisionmakers would ideally have the ability to know the dose rates that individual people are exposed to at any given moment, as well as the cumulative dose to which they have been exposed, so as to guide responders away from danger zones and ensure no one is exposed to excessive radiation. They also need to know how much radiation is being emitted in a local area, and whether this radiation is being emitted from airborne particles, water sources, artificial surfaces, soil, plants, or animals. It is important to characterize the type of contamination. For example, superficial radioactive dust can be removed more readily than deeply embedded particles or radionuclides dissolved in groundwater. Also, the handling of radioactive material may depend on the half-lives of the radionuclides it contains; short-lived radionuclides may be allowed to decay *in situ*, while longer-lived ones may require removal and encapsulation.

In doing this type of assessment, there are several key challenges. First, given varying radiation levels over time and space, very large numbers of cost-effective, durable sensors would need to be employed. A second challenge is getting sensors to where they are needed, in the numbers in which they are needed, moving them around within the affected areas, and enabling them to communicate the information

---

S. Valenovà, Z. Vavrikovà, and T. Vanek, "(137)Cs and (90)Sr Uptake by Sunflower Cultivated Under Hydroponic Conditions," *Journal of Environmental Radioactivity*, Vol. 88, No. 3, April 21, 2006.

they are collecting to analysts. All this must be done in an austere environment that has recently experienced some type of disaster.

Below, we discuss experiences in these areas with respect to Fukushima.

The Fukushima Experience with Characterizing the Extent of Contamination

The Japanese authorities[2] and U.S. military[3] conducted a number of aerial surveys for wide-area characterization of the extent of radiation being emitted in the environment, particularly in the first days and weeks following the accident.[4] These surveys provided valuable situational awareness regarding the overall extent of contamination as radioactive material dispersed initially. Aerial surveys were much faster than ground surveys; they also required fewer resources and personnel to cover a given area, while requiring neither considerable exposure of personnel nor a large number of capable, radiation-hardened ground vehicles. This was particularly critical during the period soon after the accident, when the austere environment precluded the ability to effectively deploy considerable resources for this mission.

In general, the aerial surveys provided enough fidelity over wide areas to be able to make broad decisions regarding response actions, including how to conduct follow-on ground surveillance. Given the scale of the response, they were a necessary part of it. However, aerial surveys have limited granularity in characterizing local variations in

---

[2] Junichiro Ishida, "Response to TEPCO's Fukushima-Daiichi NPS Accident and Decontamination in Off-Limits Zones," briefing, Japan Atomic Energy Agency, January 18, 2012.

[3] Andrew Feickert and Emma Chanlett-Avery, *Japan 2011 Earthquake: U.S. Department of Defense (DOD) Response*, Washington, D.C.: Congressional Research Service, R41690, June 2, 2011.

[4] Multiple Japanese agencies, the international community, and non-governmental organizations continue to conduct surveys. These survey results are neither reported nor managed by a single entity to provide coordination; however, the International Atomic Energy Agency maintains an extensive database of measurements, called the Fukushima Monitoring Database, which can be found at https://iec.iaea.org/fmd/.

radiation emissions, which is one reason why additional ground-level surveys were necessary. Moreover, Japan's mountainous topography diminished the accuracy of aerially collected data: elevated areas were closer to the heights of low-flying aircraft, so the aircraft perceived these more proximate areas as emitting higher rates of radioactivity than other, low-lying areas.

Many of the challenges associated with ground-level radiation detection were exacerbated by the effects of the earthquake and tsunami, which had damaged the infrastructure throughout the area. Radiation monitoring devices in the area had been destroyed or washed away; the only monitoring devices still in service were far from the scene, but these quickly ran out of backup power supplies. The loss of electrical power and communications infrastructure limited the ability to conduct ground-level operations, either using personnel or robots. There were also shortages of skilled personnel in the area to conduct surveys themselves or to support robots doing the surveys. Degraded transportation infrastructure, shortages of basic supplies such as potable water, and the risk to personnel all contributed to the lack of personnel for this purpose. Moreover, resources and personnel were strained by the need to re-survey areas repeatedly over time, reflecting the fact that radiation levels change due to wind conditions, precipitation, biological concentration, and human activities.

A number of extreme conditions, such as high temperatures and a lack of light, further complicated sensing. In addition, improvised approaches had to be found to characterize underwater radioactive debris, or the degree to which water itself had been contaminated. One of our sources suggested that unmanned underwater vehicles would have been useful in this context.

In multiple environments, tracing leaks and finding "hot spots"— areas emitting unusually high quantities of radiation—were both difficult. It would have been helpful to have more cognizance of the distribution of both uranium and plutonium at the site. One possible approach might be to employ naturally occurring background subatomic muons, which both plutonium and uranium "shadow" (that is, bend or stop) as the muons traverse the atmosphere and pass near the metals. By imaging the muons and detecting these shadows, a general map of likely locations of these radioactive metals could be made, much as a doctor maps human bones with X-ray shadowing.[5]

Ground robots carrying radiation sensors frequently failed, for a variety of reasons. Most were not radiation-hardened, so they degraded rapidly in high-radiation environments. (More specifically, their electronics received what was effectively misinformation from the radioactivity in which they were immersed; shielding or better circuit design could have averted this.) Robots' movements were often hampered by rubble piles, stairs, and doorknobs, as well as the tangling of their power cords. In an austere environment with numerous buildings that blocked transmissions, they had great difficulties communicating, and lacked the autonomy to operate without high-bandwidth communications. Finally, they sometimes lacked the ability or strength required to manipulate objects to perform some tasks.

Additional challenges included the analysis and dissemination of data. In many cases, data streams could not be collected from a ground robot until it had been physically recovered; further time was then required for data analysis. Where data sets were available, they were not always communicated efficiently. For example, initial radiological data from sensors in areas not washed away by the tsunami (for as long as these had power in the aftermath of the event) were not collected and integrated into real-time analysis, and were not shared until a year later. There was not a common operating picture of radioactive contamination that integrated all available information. Experts whom we subsequently interviewed cited the need to have a digital image that would be continuously and automatically updated, rather than trying to integrate handwritten notes and other disparate forms of communication. Errors in transcription were common, as might be expected. Much of the data that was gathered was not provided to governmental authorities.

---

[5]   Naturally occurring muons are scattered by heavy elements such as uranium and plutonium, and could be imaged. The difficulty with this technique is that muon detectors are relatively expensive and the muon rate is time-variable, so careful measurements of the background for comparison purposes would have to be made.

Overall, gaps in knowledge of radioactive contamination contributed to both response shortfalls and additional risk to personnel. In many cases, authorities did not know the degree to which personnel sent into the area were being exposed to radiation, or would shortly be exposed as a result of movement either by the personnel themselves or migration of radioactive material.

Potential Solutions

The need for widely dispersed sensors that can report continuously on local radioactivity and wind conditions was apparent. By providing instantaneous dose rates of different types of radiation, cumulative doses as a function of time, wind speeds, and wind directions, a widespread detection and monitoring network would aid in understanding both current and near-term future radioactive conditions. Such sensors should also be able to sense and report their locations in three dimensions, using the Global Positioning System (GPS) or similar satellite-based systems that are invulnerable to damage inflicted by Earth-bound incidents.

Such sensors can be deployed aboard unmanned aerial, ground, maritime surface, and submersible vehicles to conduct area surveys. Unmanned or inhabited vehicles—in the air, on the ground, or in the water—can also be used to widely distribute such sensors across a range of fixed locations, after which these devices will collect and transmit data continuously. (Such vehicles could also deliver other urgently needed supplies and equipment.) Personnel and inhabited vehicles could also carry such sensors with them to record local conditions, enabling themselves and others to fulfill necessary missions while remaining fully cognizant of their levels of radiation exposure. Naturally, the sensors need to be designed to be small, rugged, and low-cost to enable them to be used in all of these contexts. They would also need to be able to endure in the environment for long periods without further support, requiring some combination of batteries and the ability to "scavenge" energy from the environment (e.g., solar power, wind energy, or even energy absorbed from radiation). They would also need to be radiation-hardened.

A key aspect of this proposed solution is that the sensors need to be networked so that they can provide data in real time to analysts and modelers, who can integrate and analyze it to provide decisionmakers with actionable information. In most environments in developed countries, this would not be challenging under ordinary circumstances; numerous channels for wireless communication would enable data to be transmitted from scattered sensors to central locations. However, after a disastrous event that caused radioactive dispersal, existing infrastructure could be damaged. An effective and resilient radiation-dispersal measurement system could use some combination of unmanned vehicles and aerostats to relay information from sensors to central locations for analysis and modeling. Naturally, these would have to be deployed rapidly after an event, and would have to be distributed widely enough that sensors would not need to use a great deal of power to communicate with them. The amount of bandwidth required per sensor reading would likely be small: a few numbers to indicate the sensor's current and cumulative dose readings, as well as its location and local wind data. However, the system would need to accommodate large numbers of sensors, each emitting such data on a frequent basis. It would also need to integrate, analyze, and transmit these data sets to multiple locations where decisionmakers could use them, and where they could then prepare that information for dissemination to the general public. This information could also be sent to field locations, or to personnel in the field, to help improve their effectiveness and reduce their exposure to radiation. If bandwidth limitations precluded sending a complete data set, local gradients or contours could be sufficient to aid in decisions.

The network described above could be highly resilient: the loss of information from any given sensor would result in information shortfalls only in very localized areas. Moreover, having redundant pathways for communications—by having more relay stations than would be required if all systems were working perfectly, and potentially multiple communication channels—would help to ensure resilience.

Several of our sources recommended creating "corridors" through a contaminated area in which extremely detailed surveys were conducted. Ideally, corridors would be selected based on an initial wide-area survey that indicated where radiation levels were relatively low. Continuous monitoring of these corridors, and of alternatives to portions of them, could help to enable mobility for both personnel and machines while minimizing either exposure or the need to counter it. The corridor concept dovetails well with the idea of subsequently creating "islands" of decontamination and, eventually, pathways to them, as will be discussed in the next chapter.

Some of our sources cited the advantages of being able to conduct aerial surveys with higher degrees of granularity than were possible at the time. As noted earlier, aerial surveys have considerable advantages with respect to speed and requirements for resources and personnel in an austere environment. Several approaches to improving aerial surveys might be considered. One is the use of unmanned aircraft, some of which can fly particularly slowly and relatively close to the ground. Another is the development of highly detailed topographical electronic maps, which could be used to subtract out apparently elevated levels of radiation that were actually due to variations in altitude. Some of our sources suggested that aerial surveys are inherently limited in terms of granularity and accuracy due to wind conditions and other atmospheric phenomena, and that a better use of either inhabited or unmanned aircraft is to drop sensors to the ground and subsequently serve as communications relays for those sensors.

A number of solutions could help to improve the performance of ground robots. Radiation-hardening through shielding and redundancy in circuit design could enable greater endurance to highly radioactive material. Improvements in both mobility and autonomy are needed; such advances would be beneficial in a variety of industrial, military, and consumer contexts, as well as in disasters of any kind. We will discuss these issues at greater length in the chapter on robotics.

Chapter Findings

Rapidly deployable sensors capable of surveying large areas quickly are critical for both initial characterization and on-going monitoring of a radioactive dispersal event. In addition, more finely grained local sensors suitable to support the establishment and maintenance of safe corridors and staging areas along with hardened unmanned sensor-carrying systems would be needed. Gaps in knowledge about the extent of contamination early in the Fukushima disaster prevented fully effective responses by officials. Thus, information technologies to quickly and accurately share and display sensors' radiological measurements in real time are needed to support disaster response.

CHAPTER THREE

Preventing Radiation Damage and Further Dispersion of Material

In this chapter, we review five different aspects of mitigating radiation dose to and its effects on individuals and the further dispersion of contaminated material:

- use of radiation suits and collective protection
- personal dosimetry
- medical, biological, and genetic response
- minimizing contamination via agriculture, food, and drinking water
- containing contaminated materials to prevent further dispersion.

The Problem of Radiation Suits and Collective Protection

Personnel who need to operate in contaminated environments obviously require protection appropriate to the amount and types of radiation in their vicinity, as well as the period of time they will spend in the environment. Typically, personnel will wear full-body radiation suits engineered to provide a radiation barrier, along with self-contained breathing units, providing complete environmental separation—much as a space suit would. These suits provide good radiation protection, but are also heavy and not usually climate-controlled. At Fukushima, sources noted that the existing suits caused heat fatigue and intense headaches, making personnel less capable and reducing the amount of time they could wear the suits. Another problem was that the extreme temperatures at the plant sometimes melted the radiation suits' boots.

Potential Solutions

There are several approaches to making better suits that allow users to operate more effectively and safely. One is the use of exoskeletons—suits that are more like external armor than clothing—that can be designed to enhance human strength, rather than sap it. Another is to find more flexible (or otherwise less encumbering) materials that could serve the same function as the existing suits. This may be fundamentally impossible, since the radiation-attenuating properties of lead and other shielding materials is correlated with their density, but it could be pursued.

Collective protection could be a replacement for, or a supplement to, individual protection. Rapidly deployable barriers that would protect personnel from external radiation could minimize the amount of gear that they would be required to wear within the shelter. For example, such a shelter could be set up as an "island" in a less contaminated area surrounded by more contaminated ones, then used as a base for launching, controlling, and supporting robots. Conversely, if there were particular small, intense "hot spots" that were irradiating a wider area around them, these could be encapsulated by deployable barriers to reduce the exposure of personnel outside them.

Another alternative would be to remove people from the environment altogether, replacing them with robots that could be controlled from afar. However, completely eliminating personnel from such environments would require huge advances in robot autonomy, local support links that the robots could use (e.g., for recharging), radiation hardening for work in the most hazardous areas, and extensive, high-bandwidth communication links. It might also contribute to some loss of situational awareness if people were physically separated from the scene.

Some of our sources suggested having a readily deployable cache of radiation suits and personal dosimeters for use in an emergency.

The Problem of Personal Dosimetry

Personnel in and around suspected radiation sites require real-time monitoring of external and internal radiation exposure to minimize personal risk of illness and to know when to seek treatment. Historically, technological limitations made personal dosimetry—the monitoring of individuals' radiation exposure—complex, error-prone, and expensive. In the past, dosimetric instruments were difficult to calibrate and only operators with considerable training could use them effectively. Furthermore, these instruments rarely collected any kind of geographic or temporal data to contextualize radiation readings. As a result, personal dosimetry of all kinds was fraught with uncertainty. Personal dosimetry of large populations' external exposures was practically impossible, necessitating the substitution of dose reconstructions of questionable accuracy. For instance, after the Chernobyl disaster, despite sizable stockpiles of dosimetric instruments gathered as part of its civil defense program, the Soviet Union struggled to monitor the external radiation exposures of emergency workers, and lacked the resources to do the same for its citizens living in contaminated areas. Furthermore, although external dosage to individuals can, in principle, be measured, thus giving the basis for estimating health risk, *internal dosimetry*—that is, measuring radiation dosage due to ingested radioactive material—cannot be measured directly today. At Fukushima Dai-Ichi, individual internal dosage was estimated by measuring whole-body counts, building human-tissue models, and estimating individual exposure.[1] External radiation exposure was measured using wearable personal dosimetry devices. These devices typically provided alarms in high-radiation areas, but not always real-time cumulative dosages to the wearer, nor real-time integration of the measurements with information technology to create contextual "maps" of the radiation environment for use by personnel.

---

[1]   Makoto Miyazaki, Akira Ohtsuru, and Tetsuo Ishikawa, "An Overview of Internal Dose Estimation Using Whole-Body Counters in Fukushima Prefecture," *Fukushima Journal of Medical Science*, Vol. 60, No. 1, 2014, pp. 95–100.

Potential Solutions

Fortunately, advances in both dosimetric instruments and information technology promise to overcome the historical limitations of personal dosimetry, opening a wide range of possibilities that the Japanese have begun to exploit in the aftermath of the Fukushima Dai-Ichi accident. Integration of dosimetric instruments with digital technology is not only making these devices cheaper and easier to use, but also making fine-grained monitoring of populations' radiation exposures feasible. Inexpensive GPS and wireless data transmission technologies could be incorporated into personal dosimeters, collecting a wealth of data not only about individuals' external radiation doses, but also about where and when they received these doses. Multiple interview subjects expressed enthusiasm for these approaches, and Japan's Nuclear Regulation Authority (NRA) has announced its intention that citizens returning to evacuated areas wear personal dosimeters. Trial distribution of personal dosimeters has already begun, but many of the devices provided to citizens are older designs that can only register doses accumulated over a period of several months. The NRA is developing a more advanced model that can inform users of exposure on an hour-to-hour basis, allowing them to identify higher-radiation areas. While Japanese researchers are crafting personal dosimeters integrated with GPS receivers and smartphones, interview subjects warned that these devices might not achieve the public acceptance required for widespread adoption due to privacy concerns.

Educational and psychological challenges complicated the adoption of personal dosimetry for exposed populations. Citizens often lack an understanding of the meaning of radiation readings and are confused by the figures and units of radiation exposure, and one interview subject expressed his concern that the public would not trust the instruments' accuracy. As part of its program to distribute personal dosimeters, the NRA foresees the establishment of a staff of facilitators who will, among other roles, help members of the public understand and interpret their dosimetric results. To make the dosimeters easier to use, an interviewee proposed the employment of a simple color-coded display (red = "high," green = "low"). Several interviewees suggested that personal dosimeters should include features commonly associated with other radiation survey instruments, such as real-time radiation level readouts and alarms, in part due to a belief that this information will help assuage popular concerns. Without an effective education campaign to help citizens make sense of these readings, however, this additional information may merely further stoke public fears of radiation hazards.

Personal dosimeters collecting real-time geospatial data about radiation fields could be of significant utility to U.S. military personnel and emergency responders. These devices require no technological breakthroughs to become feasible, but would benefit from advances reducing their still-considerable upfront cost. High-granularity, up-to-date information about radiation hazards and personal exposures could then be analyzed to identify hot spots and make informed decisions about decontamination and exclusion, enabling more cost-effective management of these problems. Crafting effective software to analyze this data might prove just as or more difficult than developing the dosimeters themselves.

In addition to external dosimetry, Japanese researchers are also seeking improved methods for monitoring the internal radiation exposures of Japanese citizens. To determine the radiation doses received from internal body burdens of radioisotopes such as cesium-137 and strontium-90, the Japanese are employing whole-body counters, but current devices have proven inadequate for the specific demands of Japanese citizens. In particular, many citizens want to examine the radiological burden of children and infants, but most whole-body counters are designed for adults. Not only do children's relatively lower masses pose difficulties for whole-body counters, the machines also typically require patients to remain still for lengthy periods for accurate results, which may be unrealistic to expect from the very young. One interviewee noted that the considerably quicker FASTSCAN machine, which can scan an adult in an average of one minute, is not designed to examine children. Furthermore, most existing whole-body counters are only capable of detecting relatively high-energy gamma emitters, such as cesium-137, and not beta emitters, such as strontium-90. While detecting radio-strontium in living people is difficult, it is technically possible thanks to the low-energy gamma rays produced by bremsstrahlung of beta particles. In the 1970s, Soviet researchers exploited this principle to construct a whole-body counter called SICH-9.1, which they used to measure the radio-strontium body burden in populations surrounding the Mayak weapons complex. Although technically crude and physically large, the success of this device suggests that with improved detector technology a faster, cheaper, and more accurate whole-body counter could be developed capable of detecting both cesium-137 and strontium-90. Inexpensive, sensitive, and reliable detectors for low-energy gamma rays offering a high degree of position sensitivity would be essential for such a device, and could be a target for DoD research efforts.

The Problem of Medical and Genetic Aspects of Radiation Health Effects

Anecdotal evidence from populations living in areas with a high radiation background, atomic bomb survivors, and nuclear industry workers has convinced some in the health physics community that there exists considerable variation between individuals' sensitivity to radiation, and that if the basis of this phenomenon could be understood it might be exploited to manage the effects of radiation emergencies. For instance, there are populations living in an area of Iran with tens or hundreds of times higher radiation background than the global average due to rich thorium deposits, but the prevalence of radiation-related diseases in these areas is not significantly higher than average. Furthermore, it is known that the BRCA1 and BRCA2 genes, mutated versions of which result in higher rates of breast cancer, are in fact associated with a DNA repair mechanism that responds to radiation damage to cells. Therefore, individuals carrying damaged versions of these genes are more susceptible to radiogenic breast cancers. Although the hypothesis that radio-sensitivity varies substantially within populations is far from universally accepted, one interview subject echoed it, and pointed to in vitro studies on cell cultures supporting it. This view also finds considerable support among researchers in the former Soviet Union studying the health impact of the Chernobyl disaster.

Potential Solutions

If significant genetic variations exist in individuals' sensitivity to the health effects of radiation, a range of possible technological interventions could be developed to reduce the health costs of radiation exposures. For instance, if genetic variations associated with higher radio-sensitivity could be identified, a genetic screening program could be enacted to prevent particularly radio-sensitive individuals from living or working in contaminated areas. More ambitiously, either gene therapies or new radio-protective drugs could be developed to reduce individuals' vulnerability to the health effects of radiation. Substantial advances in the study of the biomolecular mechanisms underlying cells' response to radiation exposure may be necessary to determine the feasibility of these approaches, but even marginal reductions in radio-sensitivity could prove extremely useful. Reducing individuals' sensitivity to radiation, however challenging, could not only radically reduce the costs of radiological disasters, but also vastly increase the ability of the U.S. military to operate in hostile radiation environments, making it an alluring objective for DoD.

Regenerative medicine provides another potentially promising area of investigation to ameliorate the health effects of radiation exposure. Interview subjects suggested that stem cell research might prove a fruitful avenue of research to this end, proposing, for instance, that the availability of regenerative medicine would increase the willingness of individuals to work in contaminated areas. Exploiting this technique would likely require collecting and storing individuals' stem cells prior to radiation exposure so that these might be used as needed for subsequent stem cell treatments.

The Problem of Agriculture, Food, and Drinking Water

At Fukushima, inhalational and external exposure to radiation were most acute shortly after the accident. Soon afterwards, ingestion of radioactivity became the primary risk to human health (outside of the most irradiated areas, such as the plant itself). The earthquake and tsunami's disruption to local food supplies actually helped to prevent people from consuming irradiated material: local food was not available, so individuals were consuming food from elsewhere.

Soon after the event, milk and vegetables became particular concerns. Cows consumed radioactive material, excreting much of it in their milk (flesh absorbed little of the radioactive material). Mushrooms were particularly prone to absorb radioisotopes such as cesium-137, while root vegetables absorbed very little. Perhaps surprisingly, there was no correlation between cesium levels in the local soil and radioactive cesium in rice. Every bag of rice grown in the area was tested for radiation using an assembly line system, but only a handful of bags were found to exceed conservative radioactivity limits for human consumption, even though much of the rice was grown in contaminated soil.

Fish were contaminated both by the initial event and by continuing radioactive runoff into streams and the sea. However, certain cephalopods and crustaceans (notably octopi, squid, and crabs) have been absorbing less radioactive material than other sea creatures. Fish are now being assessed using a non-destructive testing regimen developed by Tohoku University. The challenges, as in so many areas of response, are scale and cost. Monitoring 100 percent of foodstuffs would restore confidence and enable the local economy to regenerate, but the expense of doing so for so many types of food may be prohibitive.

The Japanese government encouraged people to consume apple pectin jelly to prevent absorption of radiation. However, this had little impact, because people were generally consuming very little radioactive material. Relatively soon after the accident, very little of the food being produced in Fukushima posed any measurable risk; however, the psychological effects linger.

Potential Solutions

One approach to countering many plants' uptake of radioactive cesium is to use especially potassium-rich fertilizers. Plants that uptake cesium do so because it is chemically similar to the potassium that they are trying to consume; suffusing their environments with potassium can make it prevalent enough that they will absorb very little cesium. In a similar vein, livestock can be fed absorptive materials that will help to keep radionuclides from being absorbed into their bodies, enabling the animals to simply excrete them.

Some of our sources noted that dietary advice could be targeted to individuals based on their health, age, and other personal characteristics. For example, children might be advised not to eat mushrooms.

The Problem of Containing Contaminated Materials to Prevent Further Dispersion

Effectively containing contaminated materials—preventing them from further dispersing in the environment—can help greatly to reduce both exposure and cleanup costs. In the aftermath of an incident, this will likely take several forms:

- preventing further dispersion of radioactive material from the source of the event
- preventing contaminated dust from becoming airborne
- preventing contaminated water from entering the environment (or propagating further throughout the environment).

The first of these is beyond the scope of our study and is really only applicable when a nuclear power plant, rather than a nuclear weapon or radiation dispersal device, is the cause of the incident. In this section, we explore the other two aspects of containment of material, namely preventing contaminated dust from becoming aerosolized and preventing contaminated water from entering the environment. (We will describe the disposal of contaminated materials that have been collected in Chapter Five.)

Potential Solutions
Preventing contaminated dust from becoming aerosolized. Dust can be aerosolized both by natural forces, such as winds, and by human activity, such as digging or driving. At Fukushima, one of the ways of preventing the dispersion of dust was extremely simple: spraying water

on it. In some cases, longer-lasting sprays and gels were also used for this purpose.

Preventing contaminated water from entering the environment. As one of our sources noted, one of the differences between the Fukushima and Chernobyl events was the abundance of water at Fukushima: it is a much wetter environment than northwestern Ukraine. Moreover, water, including seawater, has been used in vast quantities at Fukushima to try to cool down the facility. In this context, a huge fraction of the overall resources aimed at responding to the incident have been concentrated on dealing with two issues involving the containment of water. The first was preventing water that had already been contaminated from entering the wider environment, and the second was interrupting the flow of groundwater that would otherwise seep into heavily contaminated areas, become radioactive, and then re-enter the environment.

The contaminated seawater used for cooling the facility has been contained within tanks, but the seawater's corrosive properties and radiation are damaging portions of the tanks. Leakage is an unavoidable problem. In addition, the tanks remain vulnerable to earthquakes and other extreme events.

At one point, TEPCO was planning to construct an "ice wall" buried in the ground around the Fukushima facility to prevent the flow of groundwater; specifically, an array of tubes containing coolant will freeze the ground around the perimeter of the facility creating an underground wall. One of the concerns regarding the ice wall is that if it prevents water inflow (and outflow), a loss of groundwater pressure may accelerate the rate at which heavily contaminated water leaches into the environment from the plant through gaps in the ice wall.

Chapter Findings

Better individual monitoring and protection, such as improved hazmat suits, personal dosimetry, and personalized medical approaches to radiation hazards for humans, is needed to ensure worker and resident safety. The lack of such technologies contributed to the negative public perceptions and fear about the event. Approaches to preventing land agricultural uptake of cesium and strontium exist today, facilitating the safety of locally grown food. However, prevention of sea-life contamination remains difficult. Public perceptions about the safety of local food motivates technological development in this discipline. Finally, dust-suppression methods in the local area were effective, but large-scale water management remains challenging.

CHAPTER FOUR

Decontamination and Collection of Radioactive Material

One of the most challenging aspects of responding to an incident like that at Fukushima is decontaminating the environment to a sufficient extent that human activities can resume. At Fukushima, widespread cesium contamination persists in the soil, plants, and urban areas surrounding the plant; at the plant itself, water also is contaminated. This variety of contaminated material complicates the decontamination approach. However, broadly speaking, there are three approaches to decontamination:

- physical decontamination—including removing superficial or airborne radionuclides by applying direct mechanical force and/or using flowing water or air
- chemical decontamination—which takes advantage of atomic-level interactions to concentrate radionuclides of particular elements into a smaller mass of material
- biological decontamination—which employs living systems' preferences for absorbing the atoms of particular elements to concentrate radionuclides from a large medium (such as soil or water) into a smaller mass of organic material (such as a sunflower or mushroom).

In general, only a small percentage of the mass of waste collected is composed of radioactive material. To the degree that radioactive material can be selectively separated from non-radioactive material, a smaller quantity of radioactive waste will need to be disposed of. In general, physical processes are the least selective, while biological ones are the most selective. Reducing the mass of waste is imperative because of the need to secure the collected radioactive material for very long periods of time. Common radioactive isotopes often have long half-lives (for example, both cesium-137 and strontium-90 have half-lives of approximately 30 years); waste containing these isotopes needs to be securely encapsulated for a century before the radiation emitted falls by an order of magnitude.

Physical Methods of Decontamination

Many items used for physical nuclear decontamination in the case of a disaster are not specialized tools, but rather are selected from among easily available items. For example, in the case of Fukushima, people have been using shovels, brushes, rags, and similar household items for basic decontamination processes such as removing topsoil or grass, removing mud from gutters, and cleaning roofs.[1]

Other, more specialized equipment has also been helpful for physical decontamination. For example, HEPA (high-efficiency particulate air) filter vacuums, which are used for a number of other purposes, have been employed and adapted for nuclear abatement. High-pressure water sprayers have been used extensively on houses, sidewalks, and roads.[2] Conventional street-sweeping vehicles have also been used for decontamination. Notably, ultrahigh-pressure hoses have been useful at decontaminating paving-stone roads.[3] Once the water is discharged, another street-cleaning vehicle is used to pump up the contaminated water. For large-scale structure removal, explosives may be an expe-

---

[1] Shunsuke Kimura, "Cleanup Work Progresses in Fukushima, but Residents Still Concerned," *Asahi Shinbun*, March 7, 2013. For a good reference on the success of all sorts of readily available processes for decontaminating environments, such as man-made surfaces, housing environments, forests, rural areas, and agricultural areas, see Jørn Roed, K. G. Andersson, and H. Prip, *Practical Means for Decontamination 9 Years After a Nuclear Accident*, Roskilde, Denmark: Risø National Laboratory, 1995.

[2] Jeffrey Hays, "Clean Up, Decontamination and Radioactive Debris and Soil Around Fukushima," *Facts and Details*, April 2012.

[3] Hays, 2012.

ditious method, but are problematic because of the potential dispersion of contaminants. Concrete shavers have been adapted for disaster applications where many large concrete surfaces need to be decontaminated, and have the advantage of being relatively low-cost. Additionally, "scabblers," or machines that grind surfaces, have been employed for decontamination.[4]

Chemical Methods of Decontamination

The experiences of various countries' nuclear industries, as well as from the U.S. Environmental Protection Agency Superfund program, indicate that several chemical decontamination methods and electrochemical processes exist for remediation purposes. The chemical processes involve spraying liquids on affected sites; applying foams or gels to affected materials; or bathing affected items in chemical vats. Preferred chemical means of decontamination depend on the characteristics of the contamination; the size, quantity, and composition of items to be decontaminated; and the experience of the administrator, since substantial expertise is required when using chemical decontamination, especially when using technology in which organic chemicals bind to metallic ions, some of which are radioactive.[5] A key disadvantage is that there are storage challenges associated with handling the contaminated liquid waste that results from these processes. Many chemical methods are only recommended for closed systems, and therefore are less applicable for open-area disaster situations.

---

[4] Sang Don Lee, Robert L. Sindelar, and Mark B. Triplett, *Report of the United States Embassy Science Fellows Support to the Government of Japan—Ministry of the Environment: Observations and Commentary on Remediation of the Lands Off-Site from the Fukushima Daiichi Reactors*, U.S. Environmental Protection Agency, SRNL-RP-2013-00303, July 2013, pp. 56–60.

[5] E. Feltcorn, *Technology Reference Guide for Radioactively Contaminated Media*, U.S. Environmental Protection Agency, EPA 402-R-07-004, October 2007.

Decontamination of Water

Decontamination of water is a particularly useful context in which to apply both chemical and physical methods of remediation. There are three distinct types of radioactivity associated with water. The easiest to remedy is that emitted by particles suspended in water without being chemically dissolved by it. Such particles can be mechanically filtered out, or simply allowed to settle to the bottom of tanks and then mechanically removed.

A slightly more challenging problem is posed by radionuclides that are dissolved in water. For example, radioactive cesium-137 will readily dissolve in water, meaning that individual atoms of the material become interspersed with the water molecules (each losing an electron to become positive ions as they do so). Radioactive dissolved ions can be removed from water by adding a substance that will cause the cesium to precipitate out of solution, then mechanically filtering the solid precipitate or allowing it to settle at the bottom of the tank for removal.[6] Sandia National Laboratories has developed an inorganic molecular sieve that captures and separates specific elements from radioactive wastewater through the use of crystalline silicon-titanate.[7] Separately, Kurion Inc. has developed an Ion-Specific Media System that absorbs radioactive contaminants to separate them from water.[8] The waste may be then encased in glass, in a process called vitrification.[9] A key challenge associated with decontamination of water is the large scale at

---

[6] For example, see Harold Rogers, John Bowers, and Dianne Gates-Anderson, "An Isotope Dilution–Precipitation Process for Removing Cesium from Wastewater," *Journal of Hazardous Materials*, Vol. 243, December 2012.

[7] Sandia National Laboratories, "Sandia Labs Technology Used in Fukushima Cleanup," news release, May 29, 2012.

[8] Melissa Mahony, "Radioactive Water Cleanup Steams Ahead at Fukushima," *Smart-Planet*, August 29, 2011.

[9] The disposition of this waste remains pending as of this writing. In nuclear waste management, vitrification means the mixing of radioactive elements with melted glass (silicon) fragments, and cooling the conglomeration to create a water-resistant glass-like substance. The nuclear waste is not, strictly speaking, turned into glass, but rather tightly bound to glass preventing migration into the environment.

which it typically needs to be undertaken, involving vast quantities of chemical filters and energy.

The most difficult type of radioactive water remediation arises when water molecules contain tritium, a radioactive isotope of hydrogen with a half-life of 12 years. The chemical attributes of tritium atoms are nearly identical to those of other hydrogen atoms, making water molecules containing tritium very difficult to separate from bulk water. The primary methods for achieving such separations involve the use of molecular membranes and adsorptive beds.[10]

Biological Methods of Decontamination

Biological methods for decontamination entail using the ability of living things to concentrate specific metals, including radionuclides. For example, some plants preferentially absorb specific radionuclides, and these can be grown in a contaminated area to absorb the targeted substance, and then be collected for disposal. Sunflowers are currently being used in Fukushima to extract cesium, and will then be decomposed by hyperthermophilic aerobic acid bacteria to further reduce the mass that needs to be disposed of.[11] Such a process can drastically reduce the mass of radioactive waste that needs to be stored. Mushrooms have been suggested for their absorptive properties, as has cannabis for the speed at which it grows. It has also been found that some types of algae can be used to help decontaminate water.[12]

---

[10] See, for example, D. W. Jeppson, G. Collins, L. Furlong, and S. L. Stockinger, "Separation of Tritium from Wastewater," paper presented at *Waste Management 2000* conference, Tucson, Ariz., February 2000; and Myung W. Lee, "Method and Apparatus for Separation of Heavy and Tritiated Water," U.S. Patent No. US 6332914 B1, December 25, 2001.

[11] Hays, 2012. A key challenge is that sunflowers extract about 0.5 percent of the cesium in the soil around them per life cycle, so it would take a number of replantings of sunflowers to extract a large fraction of the cesium.

[12] Maxime Goualin, "The Japanese Nuclear Power Accident: Is Seaweed and Cannabis Being Used to Treat Nuclear Power Waste?" *Cereplast RSS*, August 11, 2011.

The Fukushima Experience

At Fukushima, the character of the environment exacerbated the difficulty of the decontamination effort. As several of our sources indicated, forest and mountain environments retain radioactive materials much better than do most urban environments due to cesium capture attributes of the natural clay in soil. Cesium binds tightly to the clay in Fukushima's soil, making it difficult to extract using decontamination methods but allowing it to slowly percolate through the ecosystem and hydrological cycle. The dynamic forest environment shuffles radioactive material around from soil to the tree canopy, and then back into the forest floor or creeks as the leaves fall late in the year. Rain also causes radioactive material to shift from elevated areas to lower ones, and Japan's abundance of underground water helps to redistribute radiation. The concern is that relatively safe areas to use (or to fish in) may become more dangerous at a later date.

Decontamination efforts at Fukushima have been focused primarily on the extensive use of simple technologies, rather than the selective use of more advanced ones. Most of the decontamination effort has consisted of gathering topsoil in bags, cleaning surfaces with rags and hoses, scraping off the external surfaces of concrete, and other low-technology approaches. These types of processes are labor-intensive, making them economically costly and dangerous due to the exposure of personnel; as one of our sources pointed out, occupational health issues due to radiation exposure may be more important than public health issues stemming from it over the long term. Decontaminated locations then need to be monitored or re-surveyed to ascertain the degree to which contamination has returned, requiring still more labor. These processes also generate large quantities of contaminated water and solid materials that require disposal.

At Fukushima, radioactive materials were often found to collect in relatively inaccessible hot spots from which they did not disperse, such as cracks and corners. The result is that there was an ongoing need to scrape those hot spots to remove the surface contamination.

The vast quantities of contaminated water at Fukushima, especially at the plant—either used for reactor cooling or simply groundwater in contact with the reactor—have not been fully decontaminated. At present, after partial treatment to remove radioactive cesium and strontium (but not radioactive tritium), water is accumulating in tanks, with no clear plans to fully decontaminate it. No known cost-effective method exists to remove the remaining radioactive tritium, which is nearly chemically identical to water.

Overall, there is simply no cost-effective way to decontaminate large areas, large quantities of water, or large amounts of human infrastructure. Small, prioritized areas that are particularly important for human usage can be decontaminated, as can areas that merit decontamination for other reasons (e.g., a small, highly radioactive area upstream of a city). However, there will always be some residual contamination that exceeds the amount present prior to the event. Addressing public concerns in this regard may prove as challenging as the decontamination effort itself.

Potential Solutions to the Decontamination Problem

In considering potential solutions to the problems of decontamination, a key consideration is that the nuclear industry has already devoted considerable resources to develop such solutions.

A number of sources suggested that the ability to vitrify radioactive material *in situ* would greatly reduce the resources required for decontamination and make it more viable over wide areas.

An alternative would be improved biological decontamination. As noted earlier, cesium accumulates in living organisms because of its chemical similarity to potassium, which organisms need and consume. Developing plants, algae, or even bacteria that would have a greater preference for cesium than existing organisms do could help in removing radiation from soil. Ideally, these organisms could also be designed to incorporate other elements whose radionuclides might also be distributed by an event, such as strontium and uranium. The organisms can then be collected and desiccated, generating much smaller quantities of more concentrated waste than would be involved in removing the topsoil.

Some of our sources suggested that a solution to the hot spot problem could be a robot with a snake-like extension that could scrape deeply in tight spaces and then vacuum or sweep up the displaced material. A particular challenge would be providing a robot with the autonomy to be able to discern precisely where such hot spots were and how to angle its gear so as to clean them.

Chapter Findings

Open-area decontamination methods for structures and land are available but labor-intensive on the large scale found at Fukushima Dai-Ichi. Open-area decontamination of water at large scale remains unsolved, although chemical methods show promise. Biological methods of decontaminating agricultural areas show promise, especially if they can cost-effectively reduce labor requirements for decontaminating large areas.

CHAPTER FIVE

Disposing of Contaminated Materials

Large quantities of radioactive material cannot be denatured by scalable, economically viable processes.[1] The only means of reducing the amount of radiation that such material emits is to wait for very long periods until a large fraction of its radionuclides have decayed and ceased to emit radiation. Once decontamination has been conducted to the extent that resources permit, the remaining radioactive material needs to be stored for long periods, ranging from years and decades to centuries (and in some cases, millennia). The time that radioactive material needs to be stored until it no longer poses a threat depends on the types and concentrations of radionuclides present,[2] as well as what is deemed to be an acceptable level of radioactivity for human exposure

---

[1] Small quantities of fissionable radioactive material such as uranium (but not cesium) could be subjected to additional sources of radiation that would transmute them into less harmful radionuclides or stable elements. This process is known as "nuclear burning." However, this is impractical for sample sizes of more than a fraction of a gram, requiring vast quantities of highly concentrated radiation under highly controlled circumstances. Moreover, it would not be possible to handle real-world samples in which diverse types of radionuclides would inevitably be intermingled with one another amidst much larger quantities of non-radioactive material. Attempts to irradiate such a sample would create more radionuclides than it would eliminate.

[2] Each type of radionuclide emits radiation at its own particular rate. This is typically described in terms of the half-life of the radionuclide: the time within which half of it will have emitted its radiation. At the end of one half-life, half of the original material remains. At the end of two half-lives, one-quarter of the original material remains; at the end of three half-lives, one-eighth of it remains, and so on. Half-lives of radioactive materials can vary by many orders of magnitude. For example, iodine-131 has a half-life of eight days, cesium-137 has a half-life of 30 years, and uranium-235 has a half-life of 704 million years.

or release into the environment. It may also depend on future technological advances. For example, the remediation capabilities available in the 23rd century may be far superior to those available now; because of this, there may be advantages to making disposal methods reversible, so that material can be accessed for future remediation or even use. Conversely, disposal methods also need to be secure against possible collection by curious or malevolent people, as well as durable enough to prevent accidental releases. Containment mechanisms need to be able to withstand the stunning intensity of select natural events—such as an earthquake, tsunami, or storm that pose a high risk—as well as the relentless, slower processes of weathering by water, wind, and biological activity. Ideally, the containment site should be relatively close to the source of the contaminated material, limiting the logistical requirements associated with moving that material. All parts of the process of moving material need to minimize both human exposure and releases into the environment, so simple approaches involving as few steps as possible are likely to be preferred. Naturally, reducing the mass that needs to be stored, or being able to vitrify it *in situ*, could make both transportation and storage of material easier.

To date, neither a site nor a containment mechanism for ultimate disposal of contaminated materials from Fukushima has been selected. Many thousands of tons of contaminated soil, rubble and biological material will need to be stored for decades due to the extent of contamination. (Disposal of nuclear waste has been a challenging issue in the United States, as demonstrated by the cessation of plans to store such waste at Yucca Mountain; the issues involved are intensified in a much smaller country with a very high population density and frequent earthquakes.) Finding a site for secure disposal of radioactive waste in the long term is a challenge; in the meantime, large quantities of contaminated soil, water, and other materials have been collected, and are accumulating in the vicinity of Fukushima.

Storing contaminated water, particularly the seawater being used at Fukushima, is especially challenging. The inherently corrosive properties of seawater, exacerbated by the destructive effects of radiation, have been damaging the rubber linings and other portions of the containment tanks being used at Fukushima, which contain many thousands of tons of contaminated water. While tanks composed of more durable materials could be used to overcome these problems, the cost of such tanks on the scale required would quickly become prohibitive. Moreover, any tank design remains vulnerable to extreme natural events, such as earthquakes, tsunamis, or violent storms.

Psychological and political issues can be as important as technical ones in shaping how contaminated material is ultimately handled. People who live near a proposed disposal location are naturally concerned about whether rupture of the containment vessel, or slow leaching of material via water percolation, could put them and their families at risk. They will also be cognizant of the importance of perceptions that could curtail their livelihoods, reduce the viability of their communities, diminish the value of their property, or even lead to irrational discrimination against them. Environmentally minded people will also have ample concerns about whether the radioactive material or containment vessel (which may contain lead) would affect the local ecology. Ultimately, where and how to secure radioactive materials will be political questions, with all the complications that entails.

In the absence of a suitable site for permanent disposal, a viable alternative approach in some instances is to disperse the material to the point that it will no longer pose a measurable hazard to people or the environment. This is most applicable in the context of contaminated water, which will rapidly disperse upon release into the ocean, whereas solid materials would migrate much more slowly. As noted previously, while many radioactive species can be removed from contaminated water, water molecules containing tritium are impossible to separate from bulk water at any meaningful scale, since the chemical properties of ordinary and tritium-containing water molecules are essentially identical. However, this last point may also be beneficial in the context of dilution: unlike other radionuclides, water molecules containing tritium will not preferentially accumulate in living organisms or on the bottom of the sea. Rather, they will disperse throughout the ocean to the point that radiation levels would be indistinguishable from background levels.

As with developing a permanent storage site, there are obviously considerable political issues involved in diluting contaminated water in the ocean, and these are somewhat separate from technical considerations. Local communities, fishermen, and the seafood-eating public will likely be concerned about possible contamination, even if dilution is technically sound. Transporting the water to remote portions of the ocean for release may be necessary for political reasons, even if unwarranted based on the actual risk involved. The economic effects of reluctance to eat seafood could devastate communities, regardless of whether radiation levels in seafood are at all elevated. Moreover, given that the ocean is a global asset, there may be harsh recriminations by other countries following the deliberate release of contaminated water into the sea. Such a release would provide ample propaganda for prospective opponents.

More exotic solutions have also been proposed for disposal of radioactive waste, though these are unlikely to succeed for both technical and political reasons. Dumping material beneath the seabed would be technically challenging and expensive, with some potential for leakage and bioaccumulation; it would also garner intense political opposition. An extreme way of removing contaminated material from the environment would be to launch it into space. However, this would be both prohibitively expensive and very risky; any type of launch failure could disperse radiation over large areas.

Chapter Findings

The large scale of contaminated material—many thousands of tons of dirt, debris, and water—preclude easy isolation from the general population. Unfortunately, no known method exists to accelerate radioactive decay at this large scale, so the material must be isolated and stored. Nuclear burning to accelerate decay could be investigated by DoD as a potential future technology, but particular attention should be given to its scaling potential. In addition, public concern about local storage of nuclear-contaminated material will powerfully shape the choice of technological solutions, so developers should consider the public acceptance of such technologies before embarking on an extensive program of work.

CHAPTER SIX

Robotics Issues

Robots played two crucial roles at Fukushima: they provided valuable reconnaissance data about the environment (including radiation levels inside and outside of buildings) and were used to help manipulate reactor controls. In select cases, they also contributed to decontamination of indoor surfaces. However, several key issues emerged in the course of using them, some of which have been mentioned previously:

- **Communications and autonomy.** Robots were operating in an austere communications environment due to the disaster; wireless communications were further impeded by building walls and other obstructions. Coordination of communication frequencies was sometimes problematic. Robots lacked the autonomy to maneuver and respond without continual guidance, which required that they be able to send high-bandwidth video back to their controllers.
- **Mobility, dexterity, and strength limitations.** Robots were impeded by debris from the disaster, natural obstacles, and such human devices as doorknobs and stairs. They also had difficulty penetrating tight spaces within buildings. In some cases, they lacked either the dexterity or the strength to perform particular tasks.
- **Power and support requirements.** Unplugged robots had limited mission durations, while robots that were charging continuously throughout their missions had very limited ranges, as well as problems with tangling cords. Robots needed to be supported and controlled from relatively close ranges, so personnel needed

to be positioned in shelters near where the robots were used. This not only put the personnel at some risk, but also required logistical support for the personnel.

- **Lack of radiation-hardening.** Anecdotal reports indicated that the sensitive electronic components of robots were degraded by radiation when close (within meters) to highly radioactive material. Robot-mounted cameras and other sensors were also reported as susceptible to damage.

Making robots more nimble, more autonomous, less vulnerable to radiation, and with fewer support requirements would have enabled them to perform more effectively, and perhaps to take on additional roles. For example, they could be used for decontamination in more varied contexts, including outdoors and in tight spaces. They would also be more useful in other disaster contexts, as well as a wide range of other scenarios. Increasing autonomy and on-board analytical capabilities could diminish the bandwidth required for communications while also permitting robots to operate for longer periods with interrupted communications.

Some of these advances will be driven primarily based on requirements and efforts outside the context of disaster response. A number of academic, industrial, and military establishments are seeking to develop more-autonomous robots that can function for longer periods and do more diverse tasks with greater dexterity and precision; such robots would have applications ranging from assembly lines to home caregiving, or even participation in combat. On the other hand, several areas stand out as fairly unique to disaster response, particularly in a radioactive environment. Developing radiation-hardened robots may also be necessary, though this can also leverage advances in radiation hardening of satellite components or other niche fields. Sheathing electronics in lead, and/or installing redundant circuitry, can enable robots to function more effectively in radioactive environments. A second specialized application is decontamination using high-pressure hoses, dry-ice blasters, ball-bearing blasters, and/or vacuums; TEPCO is testing several robots with such capabilities. Finally, a third disaster-specific application is designing robots for austere environments, in which communications may be impeded and power sources may not be available for long periods. This is important for disaster response, but also for some military applications, as well. One option for consideration in this context is "energy scavenging," in which a robot collects energy from its environment, which it then transforms into electricity. Energy sources for scavenging include solar or wind power, and possibly radiation itself.

Chapter Findings

Robots that can negotiate obstacles, operate autonomously, and withstand hazardous environments are technologies that would be quite helpful in the aftermath of a nuclear disaster such as at Fukushima.

CHAPTER SEVEN

Earlier Lessons from the Chernobyl Experience

Twenty-five years prior to the release of radiation at Fukushima, there was a much larger nuclear accident at Chernobyl, Ukraine (then part of the Soviet Union). We have briefly examined technological aspects of the Soviet response to the Chernobyl disaster to understand what has been learned since that time and to glean further insights into desirable capabilities.

The Chernobyl Experience with Characterizing the Extent of Contamination

Following the explosion of Chernobyl Nuclear Power Plant unit 4, the Soviet government faced considerable practical difficulties characterizing the extent of contamination from the damaged reactor. These challenges resulted not only from the complex and dynamic nature of the releases (which remained considerable for weeks after the accident and only concluded after the completion of the "shelter object" around the destroyed unit), but also from technological and institutional limitations. To carry out radiological surveillance, the Soviet Union relied extensively on the capabilities of the Soviet Army Chemical Troops and the Soviet Civil Defense Forces, which pressed into service equipment and procedures originally developed in anticipation of nuclear war. Available dosimetric instruments such as the DP-5 rate meter proved inappropriate for post-Chernobyl conditions because of design oversights that caused reliability and ease-of-use problems. A lack of competent personnel also hobbled radiation surveillance efforts.

There were too few uniformed Soviet Army and Civil Defense Forces dosimetrists to carry out real-time monitoring efforts throughout all of the contaminated territories, and Civil Defense Forces learned quickly that the civilian volunteer dosimetrists it had trained lacked the skills to contribute meaningfully to the surveillance effort. As a result, Soviet authorities succeeded in characterizing the extent of contamination only after considerable delays and with limited accuracy.[1]

The Chernobyl Experience with Decontamination

Following the Chernobyl accident, the Soviet government forged ambitious plans to decontaminate and repopulate contaminated areas, only to find that it could not attain these goals. As of May 1986, Soviet leaders hoped that, with an intense decontamination effort, all evacuated populations would be able to return to their homes within the foreseeable future, helping alleviate the immense embarrassment the Chernobyl accident caused the regime. The lower-than-anticipated effectiveness of available decontamination techniques, however, quickly convinced specialists that this objective could not be attained.

Physical, chemical, and biological decontamination techniques all failed to live up to expectations following the Chernobyl disaster. Some of these disappointments arose because of the qualitative differences between radioactive contamination from nuclear explosions (for which the Soviet military and civil defense had developed its decontamination procedures) and those from the Chernobyl reactor meltdown accident. Much of the radioactivity following a nuclear explosion is contained inside glassy "fallout particles,"[2] which can often be washed off of surfaces with relative ease. Radioactive contamination from Chernobyl, by contrast, proved far more intractable and complex. Radioisotopes such as cesium-137 bound themselves chemically to surfaces, and available

---

[1] Edward Geist, "Political Fallout: The Failure of Emergency Management in the Chernobyl Disaster," *Slavic Review*, Vol. 74, No. 1, Spring 2015.

[2] Samuel Glasstone and Phillip J. Dolan, *The Effects of Nuclear Weapons*, 3rd ed., Washington, D.C.: U.S. Government Printing Office, 1977.

chemical decontamination agents failed to remove them as expected. A crash program to develop improved decontamination agents eventually enjoyed some success, but these never lived up to initial hopes for the decontamination program.[3] Following practical experiments, Soviet researchers rejected biological remediation as well. According to one Soviet scientist, this research revealed "the utter non-viability" of bio-remediation as a means of ameliorating radiological contamination.[4] In the absence of practical chemical decontamination procedures, Soviet authorities resorted to physical decontamination techniques, but the crudity and labor-intensity of these methods made them practical for only limited areas—particularly, the Chernobyl Nuclear Power Plant and the areas immediately surrounding it.

To construct the "shelter object" over the destroyed reactor, Soviet authorities carried out extensive decontamination efforts around the Chernobyl Nuclear Power Plant, aiming to reduce radiation hazards to levels acceptable for emergency short-term worker exposures. The presence of ejected fuel fragments in the area around the power plant necessitated the use of heavy machinery to remove the upper layers of soil and dispose of it in marked "graves." Unfortunately, the rapid failure of the robot bulldozers built by the Cheliabinsk Tractor Factory due to radiation and thermal conditions left the Soviet response effort without a viable unmanned option for this task. The Soviet military successfully repurposed its IMR-2 armored road-building vehicle for operations around Chernobyl. Essentially a two-man bulldozer/backhoe mounted on a tank chassis, the IMR-2 offered its occupants considerable shielding against ambient radiation hazards.[5] In addition,

[3]   Staff of Civil Defense of the Ukrainian SSR, *Dezaktivatsionny raboty v khode likvidatsii avarii na Chernobylskoi atomnoi elektrostantsii* [*Decontamination Efforts in the Course of the Liquidation of the Accident at the Chernobyl Nuclear Power Plant*], Kiev: Shtab GO Ukrainskoi SSR, 1988.

[4]   R. M. Aleksakhin, and A. N. Sirotkin, "Chernobyl'skaia katastrofa i agrarnaia nauka [The Chernobyl Catastrophe and Agricultural Science]," in A. A. D'iachenko, ed., *Chernobyl'. Dolg i Muzhestvo* [*Chernobyl: Duty and Courage*], Moscow: 4-yi filial Voenizdata, 2004.

[5]   S. Paskevich and D. Vishnevskii, *Chernobyl': Realnyi Mir* [*Chernobyl: The Real World*], Moscow: Eksmo, 2011.

conventional tractors fitted with improvised radiation shielding saw considerable service moving contaminated materials at Chernobyl.

Caught without appropriate robots to assist in the Chernobyl cleanup effort, the Soviet government both purchased foreign robots and initiated a crash program to develop their own designs. The most successful of these, the STR-1, resulted from a collaborative effort of dozens of Soviet institutions and entered service in August 1986. Resembling the Soviet Lunokhod moon rovers, the STR-1 was a small, remote-controlled bulldozer with six wheels. Lofted to the roof of the damaged plant by crane and helicopter, the STR-1 pushed contaminated debris off the ledge and into special containers awaiting below. The STR-1 is reported to have operated successfully in radiation fields characterized by tens of sieverts an hour. Unfortunately, the STR-1 and other robots arrived too late and in numbers too small to make more than a marginal contribution to the Chernobyl decontamination effort. Most of the decontamination work fell to volunteer and conscript "liquidators," who were given the ironic nickname of "biorobots."[6]

Despite considerable experience and resources for dealing with radiation emergencies, technological decontamination techniques largely failed at Chernobyl. Prior to the accident, specialists failed to foresee the qualitative differences between a nuclear power plant accident and the atmospheric nuclear tests used to develop prevailing decontamination methods. Chemical and biological decontamination techniques failed to produce the hoped-for results, while physical decontamination techniques were so disruptive that they often resulted in the destruction of the areas they were intended to preserve. Ultimately, these considerations forced the Soviet Union to accept that the "Exclusion Zone" around the damaged reactor would have to remain uninhabited for decades to come.

---

6 M. I. Malenkov and A. L. Kemurdzhian, "Opyt razrabotki i ekspluatatsii robototekhnicheskogo kompleksa STR-1 pri raschistke krovel' ChAES v 1986 gody [Experience of the Development and Use of the STR-1 Robot-Technological System During the Decontamination of the Roof of the ChNPP in 1986]," in *Ekstremal'naia robototekhnika: Sb. Trudov X Muzhdunarodnoi nauchno- tekhnichekoi konferentsii* [*Extreme Robot Technology: Collection of Works from the Tenth Scientific-Technical Conference*], Saint Petersburg: TsNII RTK, 1999, pp. 48–55.

The Chernobyl Experience with *in Situ* Containment of Radioactive Contamination

In contrast to their disappointing experience with decontamination following the Chernobyl accident, Soviet researchers enjoyed much greater success in their attempts to control the migration of radioactive contamination in the environment. Employing a variety of technologies, the Soviet government succeeded in limiting the spread of contamination in terrestrial and aquatic ecosystems, as well as in crafting agricultural techniques that significantly reduced the amount of radioactivity absorbed by crops and livestock. These measures helped limit Soviet citizens' internal and external radiation exposures, reducing the human and economic costs of the disaster.

The location of the Chernobyl Nuclear Power Plant on the Pripyat River, which merges with the Dnepr before that river flows past Kyiv (Kiev), impelled the Soviet government to undertake an extensive program to stem the spread of waterborne contamination before it seriously affected the Ukrainian capital's water supply. To forestall the migration of radioactive sediment from the accident downriver, the Soviet Union constructed a system of dams over the course of 1986, which succeeded in their goal of retaining the vast majority of the contamination in the upper reaches of the river. Moscow also developed a contingency plan to supply the city of Kyiv with water from artesian wells if the need arose, but fortunately this measure never proved necessary. Ultimately, the Soviet Union found that it could manage the waterborne radioactive contamination from the Chernobyl disaster using relatively simple technologies.

In addition to forcing the evacuation of 115,000 Soviet citizens, the Chernobyl disaster also resulted in lower-level contamination of large areas that remained inhabited, forcing the Soviet Union and its successor states to seek means of reducing the agricultural impact of the accident and controlling the internal radiation exposures of the population. Shortly after the explosion of unit 4, the Soviet agriculture ministry initiated a program in pursuit of these goals. In the following years, Soviet researchers developed techniques to reduce human radiation exposures by controlling the migration of radioisotopes such as cesium-137 and strontium-90 in agricultural ecosystems. Soviet scientists introduced special fertilizer regimens that successfully limited cesium uptake by plants. They also determined that supplanting naturally occurring grass types in meadows with faster-growing varieties, along with targeted fertilizer use, could reduce the amount of cesium-137 in meadow plants by as much as a factor of ten.

Given the prominence of dairy products in the Soviet diet, limiting radiological contamination of milk posed a critical concern for Soviet authorities following Chernobyl. Beyond the pasture management techniques just mentioned, the Soviet Union determined that sorbents such as bentonite clays and ferric hexacyanoferrate could play a vital role in agriculture in contaminated areas. Not only could these sorbents be used as soil additives to prevent plants from absorbing radionuclides, Soviet veterinary researchers determined that these substances, used as a feed additive, could lower the amount of radioisotopes livestock absorbed from their food. The effectiveness of this technique varied considerably depending on the sorbent used and the form in which it was administered. Researchers found that minerals such as zeolite were much less effective than feed additives containing ferric hexacyanoferrate administered as either briquettes or boluses. Ferric hexacyanoferrate feed additives also proved much less expensive than pasture improvement techniques, making them the preferred choice for agriculture in the contaminated regions.[7]

Chapter Findings

Although astounding societal technological progress has been made in the 25-plus years since the Chernobyl disaster, many of the nuclear mitigation techniques first used by the Soviets in 1986 have changed surprisingly little.

---

[7] Aleksakhin and Sirotkin, 2004.

CHAPTER EIGHT

Conclusions and Recommendations

Several key findings emerged from this research:

- Enabling distributed, wide-area measurement of radiation over time is critical for effective response. This can be addressed by distributing a network of small radiation sensors that transmit local information to a central hub via elevated relay stations.
- Unmanned ground vehicles for environmental characterization and response need to be tailored to the needs of austere, contaminated environments. Specifically, they require improved mobility to overcome diverse types of obstacles, high degrees of autonomy due to limited communications bandwidth, the ability to deftly manipulate objects and penetrate small spaces, long dwell times in the environment, and for those systems in the most hazardous areas, radiation hardening through improved circuit design or shielding.
- Where people must venture into contaminated areas, having means of protecting them for long periods without imposing great physical strain would be valuable. One approach might be the use of "exoskeleton" suits that would shield them with an outer layer of lead while also providing them with filtered air and enhanced strength.
- More research is needed to be able to decontaminate large quantities of water and large expanses of soil or artificial surfaces. Existing methods are primarily useful for dealing with limited quantities of material, rather than the enormous amount of contaminated material that may require decontamination after an

incident. There may be value in ascertaining the cost and other trade-offs associated with different approaches to remediation.

- Public perceptions are paramount in shaping response to an incident. In some cases, technically feasible and economically affordable solutions may be precluded or hindered by public reaction to them.

Historically, governments and specialists have tended to underestimate the difficulty of decontamination following radiological incidents. Unfortunately, following the Chernobyl and Fukushima Dai-Ichi disasters, attempts at large-scale decontamination failed to live up to expectations. In both cases, ambitious plans to decontaminate and repopulate the irradiated areas were either abandoned or scaled back after considerable investment, causing immense government embarrassment. Moreover, however effective decontamination is, it is unlikely to be able to achieve radiation levels that are as low as the background levels preceding the event. Removing radiation becomes exponentially more expensive and more difficult as radiation levels get lower. In addition, ecological effects (such as bioaccumulation and cyclical movement of radiation between trees and streams) can limit the effectiveness of any decontamination effort. Wind and water can help to disperse radiation, reducing concentrations in some areas, but often raising them in other areas that people consider important.

Another key concern regarding decontamination is its effect on the people performing it. As we have seen at Fukushima and Chernobyl, the vast majority of decontamination is done by human beings who are thereby exposed to contamination, however well-protected they are. The process of decontamination reduces the threat to public health, but at the expense of creating occupational health hazards. Ideally, future advances would enable most or all of this work to be done by robots. However, the robots would still need to be controlled and supported by humans, likely from close enough range that humans would still receive some level of exposure. (Unless the robots had self-cleaning or mutual-cleaning capabilities, some of this exposure would come from contact with the robots themselves.)

Given the dire situation with respect to decontamination, the utility of other methods of containing health effects and costs must be considered. Ultimately, concerns about contamination stem from the risk that people or other living organisms will be exposed to the material, thereby experiencing negative health effects. Such exposure can be reduced, at least in the case of humans, but only at considerable cost. The best-known alternatives include decontamination, whose challenges we have already discussed, and evacuation, which involves large-scale human, political, and economic costs. Evacuation may also not be an option for the U.S. military in some scenarios, since it may need to carry out extended operations in high-radiation areas. It could be valuable to develop technologies that would allow personnel to function in a hostile radiation environment.

Fortunately, technological and scientific progress could enable the development of complementary, multi-layered strategies aimed not merely at reducing population exposures, but also at reducing the health effects of those exposures. Near-term developments could include technologies such as detailed real-time radiation monitoring, which could be used to manage individual radiation exposures more efficiently, and improved radiation-hardened electronics to allow greater automation of tasks in contaminated areas. The development of exoskeletons offering radiation shielding could facilitate human presence in seriously contaminated areas. Bio-remediation, impractical at present, might become much more useful with organisms genetically engineered to enhance their ability to concentrate radioisotopes. To better isolate radioactive materials *in situ*, nanotechnology and molecular engineering could be exploited to create more effective absorbents for agricultural, veterinary, and medical use. One potentially game-changing possibility would be to craft medical interventions ameliorating the health impacts of radiation exposures, such as advanced radio-protective drugs, or treatments designed to repair radiation damage on a cellular level. Although very technically challenging, these technologies would revolutionize the U.S. military's ability to respond to radiation incidents, and might help reduce the health and societal costs of these events to a much more manageable level.

References

Aleksakhin, R. M., and A. N. Sirotkin, "Chernobyl'skaia katastrofa i agrarnaia nauka [The Chernobyl Catastrophe and Agricultural Science]," in A. A. D'iachenko, ed., *Chernobyl'. Dolg i Muzhestvo* [*Chernobyl: Duty and Courage*], Moscow: 4-yi filial Voenizdata, 2004.

Alpeyev, Pavel, "Japan Geiger Counter Demand After Fukushima Earthquake Means Buyer Beware," *Bloomberg News*, July 14, 2011. As of November 13, 2014: http://www.bloomberg.com/news/2011-07-15/geiger-counters-sell-out-in-post-fukushima-japan.html

FEMA—*See* Federal Emergency Management Agency.

Federal Emergency Management Agency, "Nuclear/Radiological Incident Annex," June 2008. As of December 1, 2014: http://www.fema.gov/pdf/emergency/nrf/nrf_nuclearradiologicalincidentannex.pdf

Federal Emergency Management Agency, "Federal Radiological Preparedness Coordinating Committee Report," FEMA website, updated June 26, 2013. As of December 2013: https://www.fema.gov/federal-radiological-preparedness-coordinating-committee

Feickert, Andrew, and Emma Chanlett-Avery, *Japan 2011 Earthquake: U.S. Department of Defense (DOD) Response*, Washington, D.C.: Congressional Research Service, R41690, June 2, 2011.

Feltcorn, E., *Technology Reference Guide for Radioactively Contaminated Media*, U.S. Environmental Protection Agency, EPA 402-R-07-004, October 2007.

Geist, Edward, "Political Fallout: The Failure of Emergency Management in the Chernobyl Disaster," *Slavic Review*, Vol. 74, No. 1, Spring 2015, pp. 104–126.

Glasstone, Samuel, and Phillip J. Dolan, *The Effects of Nuclear Weapons*, 3rd ed., Washington, D.C.: U.S. Government Printing Office, 1977.

Goualin, Maxime, "The Japanese Nuclear Power Accident: Is Seaweed and Cannabis Being Used to Treat Nuclear Power Waste?" *Cereplast RSS*, August 11, 2011. As of August 2, 2013:
http://www.cereplast.com/the-japanese-nuclear-power-accident-is-seaweed-and-cannabis-being-used-to-treat-nuclear-power-waste/

Hays, Jeffrey, "Clean Up, Decontamination and Radioactive Debris and Soil Around Fukushima," *Facts and Details*, April 2012. As of November 12, 2014:
http://factsanddetails.com/japan/cat26/sub162/item1856.html

Ishida, Junichiro, "Response to TEPCO's Fukushima-Daiichi NPS Accident and Decontamination in Off-Limits Zones," briefing, Japan Atomic Energy Agency, January 18, 2012.

Jeppson, D. W., G. Collins, L. Furlong, and S. L. Stockinger, "Separation of Tritium from Wastewater," paper presented at *Waste Management 2000* conference, Tucson, Ariz., February 2000.

Kimura, Shunsuke, "Cleanup Work Progresses in Fukushima, but Residents Still Concerned," *Asahi Shinbun*, March 7, 2013. As of November 12, 2014:
http://ajw.asahi.com/article/0311disaster/fukushima/AJ201303070075

Lee, Myung W., "Method and Apparatus for Separation of Heavy and Tritiated Water," U.S. Patent No. US 6332914 B1, December 25, 2001.

Lee, Sang Don, Robert L. Sindelar, and Mark B. Triplett, *Report of the United States Embassy Science Fellows Support to the Government of Japan—Ministry of the Environment: Observations and Commentary on Remediation of the Lands Off-Site from the Fukushima Daiichi Reactors*, U.S. Environmental Protection Agency, SRNL-RP-2013-00303, July 2013.

Mahony, Melissa, "Radioactive Water Cleanup Steams Ahead at Fukushima," *SmartPlanet*, August 29, 2011. As of November 12, 2014:
http://www.smartplanet.com/blog/intelligent-energy/radioactive-water-cleanup-steams-ahead-at-fukushima/8325

Malenkov, M. I., and A. L. Kemurdzhian, "Opyt razrabotki i ekspluatatsii robototekhnicheskogo kompleksa STR-1 pri raschistke krovel' ChAES v 1986 gody [Experience of the Development and Use of the STR-1 Robot-Technological System During the Decontamination of the Roof of the ChNPP in 1986]," in *Ekstremal'naia robototekhnika: Sb. Trudov X Muzhdunarodnoi nauchno-tekhnichekoi konferentsii [Extreme Robot Technology: Collection of Works from the Tenth Scientific-Technical Conference]*, Saint Petersburg: TsNII RTK, 1999, pp. 48–55.

Miyazaki, Makoto, Akira Ohtsuru, and Tetsuo Ishikawa, "An Overview of Internal Dose Estimation Using Whole-Body Counters in Fukushima Prefecture," *Fukushima Journal of Medical Science*, Vol. 60, No. 1, 2014, pp. 95–100.

Paskevich, S., and D. Vishnevskii, *Chernobyl': Realnyi Mir [Chernobyl: The Real World]*, Moscow: Eksmo, 2011.

Roed, Jørn, K. G. Andersson, and H. Prip, *Practical Means for Decontamination 9 Years After a Nuclear Accident,* Roskilde, Denmark: Risø National Laboratory, 1995.

Rogers, Harold, John Bowers, and Dianne Gates-Anderson, "An Isotope Dilution–Precipitation Process for Removing Cesium from Wastewater," *Journal of Hazardous Materials,* Vol. 243, December 2012, pp. 124–129.

Sandia National Laboratories, "Sandia Labs Technology Used in Fukushima Cleanup," news release, May 29, 2012. As of August 2, 2013:
https://share.sandia.gov/news/resources/news_releases/fukushima_cleanup/

Soudek, P., S. Valenovà, Z. Vavrikovà, and T. Vanek, "(137)Cs and (90)Sr Uptake by Sunflower Cultivated Under Hydroponic Conditions," *Journal of Environmental Radioactivity,* Vol. 88, No. 3, April 21, 2006, pp. 236–250.

Staff of Civil Defense of the Ukrainian SSR, *Dezaktivatsionny raboty v khode likvidatsii avarii na Chernobylskoi atomnoi elektrostantsii* [*Decontamination Efforts in the Course of the Liquidation of the Accident at the Chernobyl Nuclear Power Plant*], Kiev: Shtab GO Ukrainskoi SSR, 1988.

U.S. Northern Command, "Joint Task Force Civil Support," website, undated. As of December 2014:
http://www.jtfcs.northcom.mil

World Health Organization, *Health Risk Assessment from the Nuclear Accident After the 2011 Great East Japan Earthquake and Tsunami,* Geneva, Switzerland, 2013.

For Further Reading

Readers curious to follow an up-to-date status of the decontamination progress at Fukushima Dai-Ichi may find the following English-language websites helpful:

- On-site remediation:
  - Tokyo Electric Power Company (http://www.tepco.co.jp/en/index-e.html)
  - International Research Institute for Nuclear Decommissioning (http://irid.or.jp/en/)
  - Ministry of Economy, Trade and Industry (http://www.meti.go.jp/english/)
- Off-site remediation:
  - Ministry of the Environment (http://www.env.go.jp/en/)
  - Japanese Atomic Energy Agency (http://www.jaea.go.jp/english/index.html)

The list below is a partial bibliography of background material offered for the interested reader. For a full bibliographic list of sources discovered in the course of this research, please contact Cynthia_Dion-Schwarz@rand.org.

"36th World Nuclear Association Annual Symposium 2011," London, United Kingdom, September 14–16, 2011.

Akabayashi, A., and Y. Hayashi, "Mandatory Evacuation of Residents During the Fukushima Nuclear Disaster: An Ethical Analysis," *Journal of Public Health (United Kingdom)*, Vol. 34, No. 3, 2012, pp. 348–351.

Akai, J., N. Nomura, S. Matsushita, H. Kudo, H. Fukuhara, S. Matsuoka, and J. Matsumoto, "Mineralogical and Geomicrobial Examination of Soil Contamination by Radioactive Cesium due to 2011 Fukushima Daiichi Nuclear Power Plant Accident," *Physics and Chemistry of the Earth*, 2013.

Akashi, M., "Fukushima Daiichi Nuclear Accident and Radiation Exposure," *Japan Medical Association Journal*, Vol. 55, No. 5, 2012, pp. 393–399.

American Nuclear Society, *Fukushima Daiichi: ANS Committee Report*, 2012.

Anderson, Christopher, "Soviet Official Admits That Robots Couldn't Handle Chernobyl Cleanup," *The Scientist*, January 20, 1990. As of November 12, 2014: http://www.the-scientist.com/?articles.view/articleNo/10861/title/ Soviet-Official-Admits-That-Robots-Couldn-t-Handle-Chernobyl-Cleanup/

Anshari, R., and Z. Su'ud, "Preliminary Analysis of Loss-of-Coolant Accident in Fukushima Nuclear Accident," *AIP Conference Proceedings*, 2012, pp. 315–327.

Anzai, K., N. Ban, T. Ozawa, and S. Tokonami, "Fukushima Daiichi Nuclear Power Plant accident: Facts, environmental contamination, possible biological effects, and countermeasures," *Journal of Clinical Biochemistry and Nutrition*, Vol. 50, No. 1, 2012, pp. 2-8.

Aoyama, Michio, Daisuke Tsumune, and Yasunori Hamajima, "Distribution of 137Cs and 134Cs in the North Pacific Ocean: Impacts of the TEPCO Fukushima-Daiichi NPP Accident," *Journal of Radioanalytical and Nuclear Chemistry*, Vol. 296, No. 1, 2013, pp. 535–539.

Arase, David M., "The Impact of 3/11 on Japan," *East Asia*, Vol. 29, No. 4, 2012, pp. 313–336.

Bandstra, M. S., K. Vetter, D. H. Chivers, T. Aucott, C. Bates, A. Coffer, J. Curtis, D. Hogan, A. Iyengar, Q. Looker, J. Miller, V. Negut, B. Plimley, N. Satterlee, L. Supic, and B. Yee, "Measurements of Fukushima Fallout by the Berkeley Radiological Air and Water Monitoring Project," Nuclear Science Symposium and Medical Imaging Conference, 2011, pp. 18–24.

Baranovs'ka, Nataliia, *Chornobyl's'ka trahediia: Narisi z istorii* [*The Chernobyl Tragedy: Sketches from History*], Kiev: Instytut istorii Ukrainy, 2011.

Beresford, N. A., and D. Copplestone, "Effects of Ionizing Radiation on Wildlife: What Knowledge Have We Gained Between the Chernobyl and Fukushima Accidents?" *Integrated Environmental Assessment and Management*, Vol. 7, No. 3, 2011, pp. 371–373.

Beresford, N. A., and B. J. Howard, "An Overview of the Transfer of Radionuclides to Farm Animals and Potential Countermeasures of Relevance to Fukushima Releases," *Integrated Environmental Assessment and Management,* Vol. 7, No. 3, 2011, pp. 382–384.

Bevelacqua, J. J., "Applicability of Health Physics Lessons Learned from the Three Mile Island Unit 2 Accident to the Fukushima Daiichi Accident," *Journal of Environmental Radioactivity,* Vol. 105, 2012, pp. 6–10.

Beyea, J., E. Lyman, and F. N. Von Hippel, "Accounting for Long-Term Doses in Worldwide Health Effects of the Fukushima Daiichi Nuclear Accident," *Energy and Environmental Science,* Vol. 6, No. 3, 2013, pp. 1042–1045.

Bird, Winifred A., "Research Initiatives: Fukushima Health Study Launched," *Environmental Health Perspectives*, Vol. 119, No. 10, October 2011, pp. A428–A429.

Bird, Winifred A., and Elizabeth Grossman, "Chemical Aftermath: Contamination and Cleanup Following the Tohoku Earthquake and Tsunami," *Environmental Health Perspectives*, Vol. 119, No. 7, 2011, pp. a290–a301.

Blankenbecler, Richard, "Radiation Worker Protection by Exposure Scheduling," *Dose-Response,* Vol. 9, No. 4, 2011, pp. 465–470.

Boing, Lawrence E., "Decommissioning of Nuclear Facilities: Decontamination Technologies," briefing slides, International Atomic Energy Agency, October 2006. As of November 12, 2014: http://www-ns.iaea.org/downloads/rw/projects/r2d2/workshop2/lectures/decontamination-technologies.pdf

Braun, J., and T. Barker, "Fukushima Daiichi Emergency Water Treatment," *Nuclear Plant Journal,* Vol. 30, No. 1, 2012, pp. 36–37.

Caffrey, J. A., K. A. Higley, A. T. Farsoni, S. Smith, and S. Menn, "Development and Deployment of an Underway Radioactive Cesium Monitor off the Japanese Coast near Fukushima Dai-ichi," *Journal of Environmental Radioactivity,* Vol. 111, 2012, pp. 120–125.

Campbell, K., "The Roles Welding Plays in Our Lives," *Welding Journal,* Vol. 90, No. 7, 2011, pp. 28–32.

Champion, D., I. Korsakissok, D. Didier, A. Mathieu, D. Quélo, J. Groell, E. Quentric, M. Tombette, J. P. Benoit, O. Saunier, V. Parache, M. Simon-Cornu, M. A. Gonze, P. Renaud, B. Cessac, E. Navarro, and A. C. Servant-Perrier, "The IRSN's Earliest Assessments of the Fukushima Accident's Consequences for the Terrestrial Environment in Japan," *Radioprotection,* Vol. 48, No. 1, 2013, pp. 11–37.

Chino, M., H. Nakayama, H. Nagai, H. Terada, G. Katata, and H. Yamazawa, "Preliminary Estimation of Release Amounts of 131I and 137 Cesium Accidentally Discharged from the Fukushima Daiichi Nuclear Power Plant into the Atmosphere," *Journal of Nuclear Science and Technology,* Vol. 48, No. 7, 2011, pp. 1129–1134.

Cleveland, G. S., "The Advisory Team for Environment, Food, and Health: Capabilities, Mission, and Initiatives," Third International Topical Meeting on Emergency Prepraredness and Response and Robotics and Remote Systems, 2011, pp. 382–395.

"Cutting-Edge Technology Aids Fukushima Cleanup," *Oil & Energy Daily,* July 11, 2013. As of August 2, 2013:
http://www.oilandenergydaily.com/2013/07/11/radball/

Cuttler, J. M., "Commentary on the Appropriate Radiation Level for Evacuations," *Dose-Response,* Vol. 10, No. 4, 2012, pp. 473–479.

Dada, E., T. Mensah, D. Rollins, L. A. Estévez, O. Shelton, and J. Harrison, "Lessons Learned from and Economic Impacts of the Fukushima, Japan Disaster," AIChE Annual Meeting, 2011.

Danielache, S. O., C. Yoshikawa, A. Priyadarshi, T. Takemura, Y. Ueno, M. H. Thiemens, and N. Yoshidai, "An Estimation of the Radioactive S-35 Emitted into the Atmospheric from the Fukushima Daiichi Nuclear Power Plant by Using a Numerical Simulation Global Transport," *Geochemical Journal,* Vol. 46, No. 4, 2012, pp. 335–339.

Dauer, L. T., P. Zanzonico, R. M. Tuttle, D. M. Quinn, and H. W. Strauss, "The Japanese Tsunami and Resulting Nuclear Emergency at the Fukushima Daiichi Power Facility: Technical, Radiologic, and Response Perspectives," *Journal of Nuclear Medicine,* Vol. 52, No. 9, 2011, pp. 1423–1432.

D'Auria, F., G. Galassi, P. Pla, and M. Adorni, "The Fukushima Event: The Outline and the Technological Background," *Science and Technology of Nuclear Installations,* Vol. 2012, 2012.

D'iachenko, A. A., *Opyt likvidatsii Chernobyl'skoi katastrofy* [*Experience of the Liquidation of the Chernobyl Catastrophe*], Moscow: Institut stategicheskoi stabil'nosti, 2004.

Dominey-Howes, Dale, and James Goff, "Tsunami Risk Management in Pacific Island Countries and Territories (PICTs): Some Issues, Challenges and Ways Forward," *Pure and Applied Geophysics,* 2012, pp. 1–17.

Estournel, C., E. Bosc, M. Bocquet, C. Ulses, P. Marsaleix, V. Winiarek, I. Osvath, C. Nguyen, T. Duhaut, F. Lyard, H. Michaud, and F. Auclair, "Assessment of the Amount of Cesium-137 Released into the Pacific Ocean After the Fukushima Accident and Analysis of Its Dispersion in Japanese Coastal Waters," *Journal of Geophysical Research-Oceans,* Vol. 117, 2012.

Fehrenbacher, Katie, "Kurion Dominates Fukushima Radioactive Water Cleanup," *GigaOM.com*, March 13, 2012. As of August 2, 2013:
http://gigaom.com/2012/03/13/
kurion-dominates-fukushima-radioactive-water-cleanup/

Fitzgerald, J., S. B. Wollner, A. A. Adalja, R. Morhard, A. Cicero, and T. V. Inglesby, "After Fukushima: Managing the Consequences of a Radiological Release," *Biosecurity and Bioterrorism,* Vol. 10, No. 2, 2012, pp. 228–236.

Fujimura, S., K. Yoshioka, T. Saito, M. Sato, Y. Sakuma, and Y. Muramatsu, "Effects of Applying Potassium, Zeolite and Vermiculite on the Radiocesium Uptake by Rice Plants Grown in Paddy Field Soils Collected from Fukushima Prefecture," *Plant Production Science,* Vol. 16, No. 2, 2013, pp. 166–170.

Fukuda, T., Y. Kino, Y. Abe, H. Yamashiro, Y. Kuwahara, H. Nihei, Y. Sano, A. Irisawa, T. Shimura, M. Fukumoto, H. Shinoda, Y. Obata, S. Saigusa, T. Sekine, and E. Isogai, "Distribution of Artificial Radionuclides in Abandoned Cattle in the Evacuation Zone of the Fukushima Daiichi Nuclear Power Plant," *PLoS ONE,* Vol. 8, No. 1, 2013.

Funabashi, Y., and K. Kitazawa, "Fukushima in Review: A Complex Disaster, a Disastrous Response," *Bulletin of the Atomic Scientists,* Vol. 68, No. 2, 2012, pp. 9–21.

Furukawa, Fumiya, Soichi Watanabe, and Toyoji Kaneko, "Excretion of Cesium and Rubidium via the Branchial Potassium-Transporting Pathway in Mozambique Tilapia," *Fisheries Science,* Vol. 78, No. 3, 2012, pp. 597–602.

Gómez Cadenas, JuanJosé, "Fukushima, or the Black Swan of Nuclear Energy," *The Nuclear Environmentalist*, Milan: Springer, 2012, pp. 161–165.

Grambow, B., and C. Poinssot, "Interactions Between Nuclear Fuel and Water at the Fukushima Daiichi Reactors," *Elements,* Vol. 8, No. 3, 2012, pp. 213–219.

Grigg, N. S., "Large-Scale Disasters: Leadership and Management Lessons," *Leadership and Management in Engineering,* Vol. 12, No. 3, 2012, pp. 97–100.

Hachisuka, A., Y. Kimura, R. Nakamura, and R. Teshima, "Study of Radiation Dose Rate in Air at Setagaya in Tokyo," *Bulletin of National Institute of Health Sciences*, No. 129, 2011, pp. 129–133.

Han, Fei, Guang-Hui Zhang, and Ping Gu, "Adsorption Kinetics and Equilibrium Modeling of Cesium on Copper Ferrocyanide," *Journal of Radioanalytical and Nuclear Chemistry,* Vol. 295, No. 1, 2013, pp. 369–377.

Hao, LeCong, Miyako Nitta, Ryoko Fujiyoshi, Takashi Sumiyoshi, and Chau Tao, "Radiocesium Fallout in Surface Soil of Tomakomai Experimental Forest in Hokkaido due to the Fukushima Nuclear Accident," *Water, Air, & Soil Pollution,* Vol. 224, No. 2, 2013, pp. 1–8.

Hashim, M., Y. Ming, and A. S. Ahmed, "Review of Severe Accident Phenomena in LWR and Related Severe Accident Analysis Codes," *Research Journal of Applied Sciences, Engineering and Technology,* Vol. 5, No. 12, 2013, pp. 3320–3335.

Hashimoto, S., S. Ugawa, K. Nanko, and K. Shichi, "The Total Amounts of Radioactively Contaminated Materials in Forests in Fukushima, Japan," *Scientific Reports,* Vol. 2, 2012.

Hayashi, M., and L. Hughes, "The Policy Responses to the Fukushima Nuclear Accident and Their Effect on Japanese Energy Security," *Energy Policy,* 2012.

Hazama, Ryuta, and Akihito Matsushima, "Measurement of Fallout with Rain in Hiroshima and Several Sites in Japan from the Fukushima Reactor Accident," *Journal of Radioanalytical and Nuclear Chemistry,* 2013, pp. 1–7.

"HEPA Filter Vacuum Cleaners," Direct Scientific website, undated. As of August 2, 2013:
http://www.drct.com/HEPA-Vacuums.html

Higaki, S., and M. Hirota, "Decontamination Efficiencies of Pot-Type Water Purifiers for 131i, 134cs and 137cs in Rainwater Contaminated During Fukushima Daiichi Nuclear Disaster," *PLoS ONE,* Vol. 7, No. 5, 2012.

Higaki, T., S. Higaki, M. Hirota, K. Akita, and S. Hasezawa, "Radionuclide Analysis on Bamboos Following the Fukushima Nuclear Accident," *PLoS ONE,* Vol. 7, No. 4, 2012.

Higashi, Tatsuya, Takashi Kudo, and Seigo Kinuya, "Radioactive Iodine (131I) Therapy for Differentiated Thyroid Cancer in Japan: Current Issues with Historical Review and Future Perspective," *Annals of Nuclear Medicine,* Vol. 26, No. 2, 2012, pp. 99–112.

Hirao, S., H. Yamazawa, and T. Nagae, "Estimation of Release Rate of Iodine-131 and Cesium-137 from the Fukushima Daiichi Nuclear Power Plant," *Journal of Nuclear Science and Technology,* Vol. 50, No. 2, 2013, pp. 139–147.

Hitchin, P., "Public Exposure in Japan," *Nuclear Engineering International,* Vol. 56, No. 682, 2011, pp. 18–20.

Honda, M. C., T. Aono, M. Aoyama, Y. Hamajima, H. Kawakami, M. Kitamura, Y. Masumoto, Y. Miyazawa, M. Takigawa, and T. Saino, "Dispersion of Artificial Caesium-134 and-137 in the Western North Pacific One Month After the Fukushima Accident," *Geochemical Journal,* Vol. 46, No. 6, 2012, pp. E1–E9.

Husqvarna, "Swedish Robotic Technology Used to Clean Fukushima Nuclear Plant," news release, October 26, 2011. As of November 12, 2014:
http://www.husqvarna.com/au/construction/company/newsroom/news-listing/swedish-robotic-technology-used-to-clean-fukushima-nuclear-plant/

Imanaka, T., S. Endo, M. Sugai, S. Ozawa, K. Shizuma, and M. Yamamoto, "Early Radiation Survey of Iitate Village, Which Was Heavily Contaminated by the Fukushima Daiichi Accident, Conducted on 28 and 29 March 2011," *Health Physics,* Vol. 102, No. 6, 2012, pp. 680–686.

Inoue, K., M. Hosoda, M. Sugino, H. Simizu, A. Akimoto, K. Hori, T. Ishikawa, S. K. Sahoo, S. Tokonami, H. Narita, and M. Fukushi, "Environmental Radiation at Izu-Oshima After the Fukushima Daiichi Nuclear Power Plant Accident," *Radiation Protection Dosimetry,* Vol. 152, No. 1-3, 2012, pp. 234–237.

Interstate Technology and Regulatory Council, *Decontamination and Decommissioning of Radiologically Contaminated Facilities*, Washington, D.C., 2009.

Ishii, H., "Research on Farming Methods for Reducing the Absorption of Radiological Materials," *Report of Research Center of Ion Beam Technology, Hosei University, Suppl No 30*, 2012, pp. 17–20.

Ishii, N., K. Tagami, H. Takata, K. Fujita, I. Kawaguchi, Y. Watanabe, and S. Uchida, "Deposition in Chiba Prefecture, Japan, of Fukushima Daiichi Nuclear Power Plant Fallout," *Health Physics,* Vol. 104, No. 2, Feb, 2013, pp. 189–194.

Isobe, T., Y. Mori, K. Takada, E. Sato, H. Sakurai, and T. Sakae, "Robust Technique Using an Imaging Plate to Detect Environmental Radioactivity," *Health Physics,* Vol. 104, No. 4, 2013, pp. 362–365.

Iwanade, Akio, Noboru Kasai, Hiroyuki Hoshina, Yuji Ueki, Seiichi Saiki, and Noriaki Seko, "Hybrid Grafted Ion Exchanger for Decontamination of Radioactive Cesium in Fukushima Prefecture and Other Contaminated Areas," *Journal of Radioanalytical and Nuclear Chemistry,* Vol. 293, No. 2, 2012, pp. 703–709.

Jang, Mee, Alan C. Perkins, and ByungIl Kim, "Dosimetry in Accidental Radiation Exposure," in Richard P. Baum, ed., *Therapeutic Nuclear Medicine*, Heidelberg, Germany: Springer-Verlag, 2014, pp. 817–833.

Kakamu, T., H. Kanda, M. Tsuji, D. Kobayashi, M. Miyake, T. Hayakawa, S. I. Katsuda, Y. Mori, T. Okouchi, A. Hazama, and T. Fukushima, "Differences in Rates of Decrease of Environmental Radiation Dose Rates by Ground Surface Property in Fukushima City After the Fukushima Daiichi Nuclear Power Plant Accident," *Health Physics,* Vol. 104, No. 1, 2013, pp. 102–107.

Kamada, N., O. Saito, S. Endo, A. Kimura, and K. Shizuma, "Radiation Doses Among Residents Living 37 km Northwest of the Fukushima Dai-ichi Nuclear Power Plant," *Journal of Environmental Radioactivity,* Vol. 110, 2012, pp. 84–89.

Kamel Boulos, Maged N., Bernd Resch, David N. Crowley, John G. Breslin, Gunho Sohn, Russ Burtner, William A. Pike, Eduardo Jezierski, and Kuo-Yu Slayer Chuang, "Crowdsourcing, Citizen Sensing and Sensor Web Technologies for Public and Environmental Health Surveillance and Crisis Management: Trends, OGC Standards and Application Examples," *International Journal of Health Geographics,* Vol. 10, No. 1, 2011, pp. 1–29.

Kameya, H., S. Hagiwara, D. Nei, Y. Kakihara, K. Kimura, U. Matsukura, S. Kawamoto, and S. Todoriki, "The Shielding of Radiation for the Detection of Radioactive Cesium in Cereal Sample by Using a NaI (T1) Scintillation Survey Meter," *Journal of the Japanese Society for Food Science and Technology-Nippon Shokuhin Kagaku Kogaku Kaishi,* Vol. 58, No. 9, 2011, pp. 464–469.

Kaneko, K., "Towards Emergency Response Humanoid Robots," paper presented at *Mecatronics REM 2012*, Paris, November 21–23, 2012, pp. 504–511.

Karpan, Nikolai, *Ot Chernobylia do Fukusimy: Dokumental'naia povest* [*From Chernobyl to Fukushima: A Documentary Account*], 2nd ed., Kiev: S. Podgornov, 2013.

Katata, Genki, Masakazu Ota, Hiroaki Terada, Masamichi Chino, Haruyasu Nagai, "Atmospheric Discharge and Dispersion of Radionuclides During the Fukushima Dai-ichi Nuclear Power Plant Accident, Part I: Source Term Estimation and Local-Scale Atmospheric Dispersion in Early Phase of the Accident," *Journal of Environmental Radioactivity,* Vol. 109, 2012, pp. 103–113.

Katata, Genki, Hiroaki Terada, Haruyasu Nagai, Masamichi Chino, "Numerical Reconstruction of High Dose Rate Zones due to the Fukushima Dai-ichi Nuclear Power Plant Accident," *Journal of Environmental Radioactivity,* Vol. 111, 2012, pp. 2–12.

Kato, H., Y. Onda, and T. Gomi, "Interception of the Fukushima Reactor Accident-Derived Cesium-137, Cesium-134 and I-131 by Coniferous Forest Canopies," *Geophysical Research Letters,* Vol. 39, October 2012.

Kawano, Y., D. Shepard, Y. Shobugawa, J. Goto, T. Suzuki, Y. Amaya, M. Oie, T. Izumikawa, H. Yoshida, Y. Katsuragi, T. Takahashi, S. Hirayama, R. Saito, and M. Naito, "A Map for the Future: Measuring Radiation Levels in Fukushima, Japan," Global Humanitarian Technology Conference, Seattle, Wash., 2012, pp. 53–58.

Kawatsuma, S., M. Fukushima, and T. Okada, "Emergency Response by Robots to Fukushima-Daiichi Accident: Summary and Lessons Learned," *Industrial Robot,* Vol. 39, No. 5, 2012, pp. 428–435.

Kitagaki, T., T. Hoshino, Y. Sambommatsu, K. Yano, M. Takeuchi, T. Igarashi, and T. Suzuki, "Fission Product Separation from Seawater by Electrocoagulation Method," *Journal of Radioanalytical and Nuclear Chemistry,* 2012, pp. 1–5.

Kobayashi, Daisuke, Masao Miyake, Takeyasu Kakamu, Masayoshi Tsuji, Yayoi Mori, Tetsuhito Fukushima, and Akihiro Hazama, "Reducing Radiation Exposure Using Commonly Available Objects," *Environmental Health and Preventive Medicine*, 2012.

Koike, H., T. Hata, and R. Kubota, "Analysis on Human and Organizational Factors Regarding Initial Responses of Shift Teams and Field Workers to the Fukushima Daiichi NPP Accident," presentation at *ANS Winter Meeting & Nuclear Technology Expo 2012*, San Diego, Calif., 2012.

Koren, Marina, "3 Robots That Braved Fukushima," *Popular Mechanics*, undated. As of August 2, 2013:
http://www.popularmechanics.com/technology/engineering/robots/3-robots-that-braved-fukushima-7223185

Korsakissok, I., A. Mathieu, and D. Didier, "Atmospheric Dispersion and Ground Deposition Induced by the Fukushima Nuclear Power Plant Accident: A Local-Scale Simulation and Sensitivity Study," *Atmospheric Environment,* Vol. 70, 2013, pp. 267–279.

Kotoku, J., K. Ishioka, H. Kanemitsu, D. Kato, N. Sasano, H. Oba, T. Kaminaga, K. Takeshita, H. Koutake, K. Toyoda, K. Kutomi, T. Haruyama, A. Watanabe, N. Sekiya, M. Sakuramachi, A. Yamamoto, Y. Ishikawa, T. Okamoto, and S. Furui, "Basic Concepts of Radiation Protection for Nuclear Disaster Relief," *Teikyo Medical Journal,* Vol. 34, No. 3, 2011, pp. 227–234.

Krisanangkura, P., T. Itthipoonthanakorn, and S. Udomsomporn, "Environmental Dose Assessment Using Ecolego: Case Study of Soil from Japan," *Journal of Radioanalytical and Nuclear Chemistry,* 2013, pp. 1–8.

Kryshev, I. I., A. I. Kryshev, and T. G. Sazykina, "Dynamics of Radiation Exposure to Marine Biota in the Area of the Fukushima NPP in March-May 2011," *Journal of Environmental Radioactivity,* Vol. 114, 2012, pp. 157–161.

Kurihara, O., K. Kanai, T. Nakagawa, C. Takada, N. Tsujimura, T. Momose, and S. Furuta, "Measurements of 131I in the Thyroids of Employees Involved in the Fukushima Daiichi Nuclear Power Station Accident," *Journal of Nuclear Science and Technology,* Vol. 50, No. 2, 2013, pp. 122–129.

Kurion, "Kurion's Ion Specific Media Is Available to Assist Fukushima Nuclear Plant Cleanup," press release, March 29, 2011. As of August 2, 2013:
http://www.kurion.com/newsroom/press-releases/kurions-ion-specific-media-is-available-to-assist-fukushima-nuclear-plant-c

Kuwahara, T., Y. Tomioka, K. Fukuda, N. Sugimura, and Y. Sakamoto, "Radiation Effect Mitigation Methods for Electronic Systems," 2012 IEEE/SICE International Symposium on System Integration (SII), Fukuoka, Japan, 2012, pp. 307–312.

Langlois, L., "IAEA Action Plan on Nuclear Safety," *Energy Strategy Reviews,* Vol. 1, No. 4, 2013, pp. 302–306.

Lavigne, O., T. Shoji, and K. Sakaguchi, "On the Corrosion BEhavior of Zircaloy-4 in Spent Fuel Pools Under Accidental Conditions," *Journal of Nuclear Materials,* Vol. 426, No. 1-3, 2012, pp. 120–125.

Leng, Yuxiao, Gang Ye, Jian Xu, Jichao Wei, Jianchen Wang, and Jing Chen, "Synthesis of New Silica Gel Adsorbent Anchored with Macrocyclic Receptors for Specific Recognition of Cesium Cation," *Journal of Sol-Gel Science and Technology,* Vol. 66, No. 3, 2013, pp. 413–421.

Levy, S., "How Would US Units Fare?" *Nuclear Engineering International,* Vol. 56, No. 688, 2011, pp. 14–17.

Lovering, Daniel, "Radioactive Robot: The Machines That Cleaned Up Three Mile Island," *Scientific American,* March 27, 2009. As of August 2, 2013: http://www.scientificamerican.com/article.cfm?id=three-mile-island-robots

Lyons, C., and D. Colton, "Aerial Measuring System in Japan," *Health Physics,* Vol. 102, No. 5, 2012, pp. 509–515.

Mabuchi, K., M. Hatch, M. P. Little, M. S. Linet, and S. L. Simon, "Risk of Thyroid Cancer After Adult Radiation Exposure: Time to Re-Assess?" *Radiation Research,* Vol. 179, No. 2, 2013, pp. 254–256.

Maguire, David, Bingrong Zhang, Amy Zhang, Lurong Zhang, and Paul Okunieff, "The Role of Mitochondrial Proteomic Analysis in Radiological Accidents and Terrorism," in William J. Welch, Fredrik Palm, Duane F. Bruley, and David K. Harrison, eds., *Oxygen Transport to Tissue XXXIV,* New York: Springer, 2013, pp. 139–145.

Maki, Norio, "Disaster Response to the 2011 Tohoku-Oki Earthquake: National Coordination, a Common Operational Picture, and Command and Control in Local Governments," *Earthquake Spectra,* Vol. 29, No. s1, 2013, pp. S369–S385.

Mallampati, S. R., Y. Mitoma, T. Okuda, S. Sakita, and M. Kakeda, "High Immobilization of Soil Cesium Using Ball Milling with Nano-Metallic Ca/CaO/NaH 2PO 4: Implications for the Remediation of Radioactive Soils," *Environmental Chemistry Letters,* Vol. 10, No. 2, 2012, pp. 201–207.

Manolopoulou, M., S. Stoulos, A. Ioannidou, E. Vagena, and C. Papastefanou, "Radiation Measurements and Radioecological Aspects of Fallout from the Fukushima Nuclear Accident," *Journal of Radioanalytical and Nuclear Chemistry,* Vol. 292, No. 1, 2012, pp. 155–159.

Maruyama, S., "Heat and Fluid Flow in Accident of Fukushima Daiichi Nuclear Power Plant, Unit 2 (Accident Scenario Based on Thermodynamic Model)," *Nihon Kikai Gakkai Ronbunshu, B Hen/Transactions of the Japan Society of Mechanical Engineers, Part B,* Vol. 78, No. 796, 2012, pp. 2127–2141.

Masashi, T., E. Hiroko, and S. Toshikazu, "Rapid Measurement of 89,90Sr Radioactivity in Rinse Water," *Health Physics,* Vol. 104, No. 3, 2013, pp. 302–312.

Matsuda, N., K. Yoshida, K. Nakashima, S. Iwatake, N. Morita, T. Ohba, T. Yusa, A. Kumagai, and A. Ohtsuru, "Initial Activities of a Radiation Emergency Medical Assistance Team to Fukushima from Nagasaki," *Radiation Measurements*, 2013.

Matsumoto, T., T. Maruoka, G. Shimoda, H. Obata, H. Kagi, K. Suzuki, K. Yamamoto, T. Mitsuguchi, K. Hagino, N. Tomioka, C. Sambandam, D. Brummer, P. M. Klaus, and P. Aggarwal, "Tritium in Japanese Precipitation Following the March 2011 Fukushima Daiichi Nuclear Plant Accident," *Science of the Total Environment,* Vol. 445-446, 2013, pp. 365–370.

Matsumura, H., K. Saito, J. Ishioka, and Y. Uwamino, "Diffusion of Radioactive Materials from Fukushima Daiichi Nuclear Power Station Obtained by Gamma-Ray Measurements on Expressways," *Transactions of the Atomic Energy Society of Japan,* Vol. 10, No. 3, 2011, pp. 152–162.

Matsunaga, T., J. Koarashi, M. Atarashi-Andoh, S. Nagao, T. Sato, and H. Nagai, "Comparison of the Vertical Distributions of Fukushima Nuclear Accident Radiocesium in Soil Before and After the First Rainy Season, with Physicochemical and Mineralogical Interpretations," *Science of the Total Environment,* Vol. 447, Mar, 2013, pp. 301–314.

Maxwell, Sherrod L., Brian K. Culligan, and Jay B. Hutchison, "Rapid Fusion Method for Determination of Plutonium Isotopes in Large Rice Samples," *Journal of Radioanalytical and Nuclear Chemistry*, 2013/04/30, 2013, pp. 1–8.

Maxwell, Sherrod L., Brian K. Culligan, and Robin C. Utsey, "Rapid Determination of Radiostrontium in Seawater Samples," *Journal of Radioanalytical and Nuclear Chemistry*, 2013, pp. 1–9.

McCann, D. G. C., A. Moore, and M. E. A. Walker, "The Public Health Implications of Water in Disasters," *World Medical and Health Policy,* Vol. 3, No. 2, 2011.

Michal, Rick, "Red Whittaker and the Robots That Helped Clean Up TMI-2," *Nuclear News*, December 2009.

Miller, C. W., "The Fukushima Radiological Emergency and Challenges Identified for Future Public Health Responses," *Health Physics,* Vol. 102, No. 5, 2012, pp. 584–588.

Min, B. I., R. Periáñez, I. G. Kim, and K. S. Suh, "Marine Dispersion Assessment of 137Cs Released from the Fukushima Nuclear Accident," *Marine Pollution Bulletin*, 2013.

Miyake, Y., H. Matsuzaki, T. Fujiwara, T. Saito, T. Yamagata, M. Honda, and Y. Muramatsu, "Isotopic Ratio of Radioactive Iodine (I-129/I-131) Released from Fukushima Daiichi NPP Accident," *Geochemical Journal,* Vol. 46, No. 4, 2012, pp. 327–333.

Morimura, N., Y. Asari, Y. Yamaguchi, K. Asanuma, C. Tase, T. Sakamoto, and T. Aruga, "Emergency/Disaster Medical Support in the Restoration Project for the Fukushima Nuclear Power Plant Accident," *Emergency Medicine Journal,* 2012.

Morino, Y., T. Ohara, and M. Nishizawa, "Atmospheric Behavior, Deposition, and Budget of Radioactive Materials from the Fukushima Daiichi Nuclear Power Plant in March 2011," *Geophysical Research Letters,* Vol. 38, No. 17, 2011.

Motooka, T., T. Sato, and M. Yamamoto, "Effect of Gamma Ray Irradiation on Deoxygenation by Hydrazine in Artificial Seawater," *Transactions of the Atomic Energy Society of Japan,* Vol. 11, No. 4, 2012, pp. 249–254.

Motooka, T., T. Sato, and M. Yamamoto, "Effect of Gamma-Ray Irradiation on the Deoxygenation of Salt-Containing Water Using Hydrazine," *Journal of Nuclear Science and Technology,* Vol. 50, No. 4, 2013, pp. 363–368.

Mrowca, B., "Removing Heat from a Reactor in Shutdown," *Mechanical Engineering,* Vol. 133, No. 5, 2011.

Murphy, Robin R., "A Decade of Rescue Robots," paper presented at 2012 IEEE/RSJ International Conference, Vilamoura, Algarve, Portugal, 2012.

Naganawa, H., N. Kumazawa, H. Saitoh, N. Yanase, H. Mitamura, T. Nagano, K. Kashima, T. Fukuda, Z. Yoshida, and S. I. Tanaka, "Removal of Radioactive Cesium from Surface Soils Solidified Using Polyion Complex: Rapid Communication for Decontamination Test at Iitate-mura in Fukushima Prefecture," *Transactions of the Atomic Energy Society of Japan,* Vol. 10, No. 4, 2011, pp. 227–234.

Nagata, T., Y. Kimura, and M. Ishii, "Use of a Geographic Information System (GIS) in the Medical Response to the Fukushima Nuclear Disaster in Japan," *Prehospital and Disaster Medicine,* Vol. 27, No. 2, 2012, pp. 213–215.

Nagatani, K., S. Kiribayashi, Y. Okada, K. Otake, K. Yoshida, S. Tadokoro, T. Nishimura, T. Yoshida, E. Koyanagi, M. Fukushima, and S. Kawatsuma, "Gamma-Ray Irradiation Test of Electric Components of Rescue Mobile Robot Quince - Toward Emergency Response to Nuclear Accident at Fukushima Daiichi Nuclear Power Station on March 2011," Proceedings of the 2011 IEEE International Symposium on Safety, Security and Rescue Robotics, Kyoto, Japan, 2011, pp. 56–60.

Nagatani, K., S. Kiribayashi, Y. Okada, K. Otake, K. Yoshida, S. Tadokoro, T. Nishimura, T. Yoshida, E. Koyanagi, M. Fukushima, and S. Kawatsuma, "Emergency Response to the Nuclear Accident at the Fukushima Daiichi Nuclear Power Plants Using Mobile Rescue Robots," *Journal of Field Robotics,* Vol. 30, No. 1, 2013, pp. 44–63.

Nagatani, K., S. Kiribayashi, Y. Okada, S. Tadokoro, T. Nishimura, T. Yoshida, E. Koyanagi, and Y. Hada, "Redesign of Rescue Mobile Robot Quince - Toward Emergency Response to the Nuclear Accident at Fukushima Daiichi Nuclear Power Station on March 2011," Proceedings of the 2011 IEEE International Symposium on Safety, Security and Rescue Robotics, Kyoto, Japan, 2011, pp. 13–18.

Nair, R. N., Faby Sunny, Manish Chopra, L. K. Sharma, V. D. Puranik, and A. K. Ghosh, "Estimation of Radioactive Leakages into the Pacific Ocean due to Fukushima Nuclear Accident," Environmental Earth Sciences, 2013, pp. 1–13.

Nakanishi, Tomoko M., and Keitaro Tanoi, eds., Agricultural Implications of the Fukushima Nuclear Accident, Tokyo: Springer, 2013, pp. 1–10.

Nakano, M., "Long-Term Impact on the Marine Environment-Simulation of the Marine Dispersion of Released Radionuclides from Fukushima-Daiichi Nuclear Power Plant and Estimation of Internal Dose from Marine Products," Atomos, Vol. 53, No. 8, 2011, pp. 29–33.

Nakano, Masanao, and Pavel P. Povinec, "Long Term Simulations of the Cesium Dispersion from the Fukushima Accident in the World Ocean," Journal of Environmental Radioactivity, Vol. 111, 2012, pp. 109–115.

Nishino, Tomoaki, Shin-ichi Tsuburaya, Takeyoshi Tanaka, Akihiko Hokugo, "Fundamental Considerations of Evacuation Behavior of Fukushima Residents in Nuclear Emergency Event Based on Questionnaire Surveys," Proceedings of the International Symposium on Engineering Lessons Learned from the 2011 Great East Japan Earthquake, Tokyo, Japan, 2012, pp. 3120–3129.

Nöggerath, J., R. J. Geller, and V. K. Gusiakov, "Fukushima: The Myth of Safety, the Reality of Geoscience," Bulletin of the Atomic Scientists, Vol. 67, No. 5, 2011, pp. 37–46.

Norio, Okada, Tao Ye, Yoshio Kajitani, Peijun Shi, and Hirokazu Tatano, "The 2011 Eastern Japan Great Earthquake Disaster: Overview and Comments," International Journal of Disaster Risk Science, Vol. 2, No. 1, 2011, pp. 34–42.

Ohnishi, T., "The Disaster at Japan's Fukushima-Daiichi Nuclear Power Plant After the March 11, 2011 Earthquake and Tsunami, and the Resulting Spread of Radioisotope Contamination," Radiation Research, Vol. 177, No. 1, 2012, pp. 1–14.

Ohno, K., S. Kawatsuma, T. Okada, E. Takeuchi, K. Higashi, and S. Tadokoro, "Robotic Control Vehicle for Measuring Radiation in Fukushima Daiichi Nuclear Power Plant," 2011 IEEE International Symposium on Safety, Security and Rescue Robotics, Kyoto, Japan, 2011, pp. 38–43.

Ohno, T., Y. Muramatsu, Y. Miura, K. Oda, N. Inagawa, H. Ogawa, A. Yamazaki, C. Toyama, and M. Sato, "Depth Profiles of Radioactive Cesium and Iodine Released from the Fukushima Daiichi Nuclear Power Plant in Different Agricultural Fields and Forests," *Geochemical Journal,* Vol. 46, No. 4, 2012, pp. 287–295.

Ohnuki, T., and N. Kozai, "Adsorption Behavior of Radioactive Cesium by Non-Mica Minerals," *Journal of Nuclear Science and Technology,* Vol. 50, No. 4, April 2013, pp. 369–375.

Omoto, A., "The Accident at TEPCO's Fukushima-Daiichi Nuclear Power Station: What Went Wrong and What Lessons Are Universal?" *Nuclear Instruments and Methods in Physics Research, Section A: Accelerators, Spectrometers, Detectors and Associated Equipment*, Vol. 731, 2013, pp. 3–7.

Oura, Y., and M. Ebihara, "Radioactivity Concentrations of I-131, Cesium-134 and Cesium-137 in River Water in the Greater Tokyo Metropolitan Area After the Fukushima Daiichi Nuclear Power Plant Accident," *Geochemical Journal,* Vol. 46, No. 4, 2012, pp. 303–309.

Parajuli, D., H. Tanaka, Y. Hakuta, K. Minami, S. Fukuda, K. Umeoka, R. Kamimura, Y. Hayashi, M. Ouchi, and T. Kawamoto, "Dealing with the Aftermath of Fukushima Daiichi Nuclear Accident: Decontamination of Radioactive Cesium Enriched Ash," *Environmental Science & Technology,* Vol. 47, No. 8, Apr, 2013, pp. 3800–3806.

Pemberton, W., R. Mena, and W. Beal, "The Role of the Consequence Management Home Team in the Fukushima Daiichi Response," *Health Physics,* Vol. 102, No. 5, 2012, pp. 549–556.

Perrow, C., "Fukushima and the Inevitability of Accidents," *Bulletin of the Atomic Scientists,* Vol. 67, No. 6, 2011, pp. 44–52.

Po, L. C. C., "Events at Unit 1," *Nuclear Engineering International,* Vol. 56, No. 684, 2011, pp. 12–15.

Pritchard, S. B., "An Envirotechnical Disaster: Nature, Technology, and Politics at Fukushima," *Environmental History,* Vol. 17, No. 2, 2012, pp. 219–243.

Qu, J. H., J. F. He, X. J. Zhang, and R. B. Wang, "Water and Food Radiation Monitor," *Hedianzixue Yu Tance Jishu/Nuclear Electronics and Detection Technology,* Vol. 32, No. 12, 2012, pp. 1467–1472.

Rana, D., T. Matsuura, M. A. Kassim, and A. F. Ismail, "Radioactive Decontamination of Water by Membrane Processes - A Review," *Desalination*, 2012.

Reed, A. L., "U.S. Doe's Response to the Fukushima Daiichi Reactor Accident: Answers and Data Products for Decision Makers," *Health Physics,* Vol. 102, No. 5, 2012, pp. 557–562.

Romanyukha, A., D. L. King, and L. K. Kennemur, "Impact of the Fukushima Nuclear Accident on Background Radiation Doses Measured by Control Dosimeters in Japan," *Health Physics,* Vol. 102, No. 5, 2012, pp. 535–541.

Saegusa, H., H. Funaki, H. Kurikami, Y. Sakamoto, and T. Tokizawa, "Design, Construction and Monitoring of Temporary Storage Facilities for Removed Contaminants," *Transactions of the Atomic Energy Society of Japan,* Vol. 12, No. 1, 2013, pp. 1–12.

Saegusa, J., H. Kurikami, R. Yasuda, K. Kurihara, S. Arai, R. Kuroki, S. Matsuhashi, T. Ozawa, H. Goto, T. Takano, H. Mitamura, T. Nagano, H. Naganawa, Z. Yoshida, H. Funaki, T. Tokizawa, and S. Nakayama, "Decontamination of Outdoor School Swimming Pools in Fukushima After the Nuclear Accident in March 2011," *Health Physics,* Vol. 104, No. 3, March 2013, pp. 243–250.

Sage, Cindy, "The Similar Effects of Low-Dose Ionizing Radiation and Non-Ionizing Radiation from Background Environmental Levels of Exposure," *The Environmentalist,* Vol. 32, No. 2, 2012, pp. 144–156.

Sakaguchi, A., A. Kadokura, P. Steier, K. Tanaka, Y. Takahashi, H. Chiga, A. Matsushima, S. Nakashima, and Y. Onda, "Isotopic Determination of U, Pu and Cesium in Environmental Waters Following the Fukushima Daiichi Nuclear Power Plant Accident," *Geochemical Journal,* Vol. 46, No. 4, 2012, pp. 355–360.

Sakama, M., Y. Nagano, T. Saze, S. Higaki, T. Kitade, N. Izawa, O. Shikino, and S. Nakayama, "Application of ICP-DRC-MS to Screening Test of Strontium and Plutonium in Environmental Samples at Fukushima," *Applied Radiation and Isotopes,* 2013.

Sakamoto, F., T. Ohnuki, N. Kozai, S. Igarashi, S. Yamasaki, Z. Yoshida, and S. Tanaka, "Local Area Distribution of Fallout Radionuclides from Fukushima Daiichi Nuclear Power Plant Determined by Autoradiography Analysis," *Transactions of the Atomic Energy Society of Japan,* Vol. 11, No. 1, 2012, pp. 1–7.

Sanami, T., S. Sasaki, K. Iijima, Y. Kishimoto, and K. Saito, "Time Variations in Dose Rate and γ Spectrum Measured at Tsukuba City, Ibaraki, due to the Accident of Fukushima Daiichi Nuclear Power Station," *Transactions of the Atomic Energy Society of Japan,* Vol. 10, No. 3, 2011, pp. 163–169.

Sarkisov, A. A., "The Phenomenon of Perception of the Nuclear Energy Hazard in Social Consciousness," *Herald of the Russian Academy of Sciences,* Vol. 82, No. 1, 2012, pp. 8–16.

Sato, I., H. Kudo, and S. Tsuda, "Removal Efficiency of Water Purifier and Adsorbent for Iodine, Cesium, Strontium, Barium and Zirconium in Drinking Water," *Journal of Toxicological Sciences,* Vol. 36, No. 6, 2011, pp. 829–834.

Schwantes, J. M., C. R. Orton, and R. A. Clark, "Analysis of a Nuclear Accident: Fission and Activation Product Releases from the Fukushima Daiichi Nuclear Facility as Remote Indicators of Source Identification, Extent of Release, and State of Damaged Spent Nuclear Fuel," *Environmental Science and Technology,* Vol. 46, No. 16, 2012, pp. 8621–8627.

Scott, B. R., "Assessing Potential Radiological Harm to Fukushima Recovery Workers," *Dose-Response,* Vol. 9, No. 3, 2011, pp. 301–312.

Shanks, A., S. Fournier, and S. Shanks, "Challenges in Determining the Isotopic Mixture for the Fukushima Daiichi Nuclear Power Plant Accident," *Health Physics,* Vol. 102, No. 5, 2012, pp. 527–534.

Shibata, Tomoyuki, Helena Solo-Gabriele, and Toshimitsu Hata, "Disaster Waste Characteristics and Radiation Distribution as a Result of the Great East Japan Earthquake," *Environmental Science & Technology,* Vol. 46, No. 7, 2012, pp. 3618–3624.

Shimura, H., K. Itoh, A. Sugiyama, S. Ichijo, M. Ichijo, F. Furuya, Y. Nakamura, K. Kitahara, K. Kobayashi, Y. Yukawa, and T. Kobayashi, "Absorption of Radionuclides from the Fukushima Nuclear Accident by a Novel Algal Strain," *PLoS ONE,* Vol. 7, No. 9, 2012.

Smolders, E., and H. Tsukada, "The Transfer of Radiocesium from Soil to Plants: Mechanisms, Data, and Perspectives for Potential Countermeasures in Japan," *Integrated Environmental Assessment and Management,* Vol. 7, No. 3, 2011, pp. 379–381.

Srinivas, C. V., R. Venkatesan, R. Baskaran, V. Rajagopal, and B. Venkatraman, "Regional Scale Atmospheric Dispersion Simulation of Accidental Releases of Radionuclides from Fukushima Dai-ichi Reactor," *Atmospheric Environment,* Vol. 61, 2012, pp. 66–84.

Sugimoto, Jun, "Current Status of Fukushima Dai'ichi Nuclear Power Plant Accident," in Takeshi Yao, ed., *Zero-Carbon Energy Kyoto 2011,* Tokyo: Springer, 2012, pp. 203–209.

Sugiyama, G., J. Nasstrom, B. Pobanz, K. Foster, M. Simpson, P. Vogt, F. Aluzzi, and S. Homann, "Atmospheric Dispersion Modeling: Challenges of the Fukushima Daiichi Response," *Health Physics,* Vol. 102, No. 5, 2012, pp. 493–508.

Sutou, S., "Fukushima Daiichi Nuclear Power Plant Disaster: Generally Applicable Implications from Measurements of Radioactive Contaminations in Some Areas of Ibaraki and Fukushima," *Genes and Environment,* Vol. 35, No. 1, 2013, pp. 1–4.

Swartz, H. M., A. B. Flood, B. B. Williams, R. H. Dong, S. G. Swarts, X. M. He, O. Grinberg, J. Sidabras, E. Demidenko, J. Gui, D. J. Gladstone, L. A. Jarvis, M. M. Kmiec, K. Kobayashi, P. N. Lesniewski, S. D. P. Marsh, T. P. Matthews, R. J. Nicolalde, P. M. Pennington, T. Raynolds, I. Salikhov, D. E. Wilcox, and B. I. Zaki, "Electron Paramagnetic Resonance Dosimetry for a Large-Scale Radiation Incident," *Health Physics,* Vol. 103, No. 3, September 2012, pp. 255–267.

Tagami, K., and S. Uchida, "Can We Remove Iodine-131 from Tap Water in Japan by Boiling? - Experimental Testing in Response to the Fukushima Daiichi Nuclear Power Plant Accident," *Chemosphere,* Vol. 84, No. 9, 2011, pp. 1282–1284.

Tagami, Keiko, and Shigeo Uchida, "Comparison of Food Processing Retention Factors of 137Cs and 40K in Vegetables," *Journal of Radioanalytical and Nuclear Chemistry,* Vol. 295, No. 3, 2013, pp. 1627–1634.

Tagami, K., S. Uchida, and N. Ishii, "Extractability of Radiocesium from Processed Green Tea Leaves with Hot Water: The First Emergent Tea Leaves Harvested After the TEPCO Fukushima Daiichi Nuclear Power Plant Accident," *Journal of Radioanalytical and Nuclear Chemistry,* Vol. 292, No. 1, 2012, pp. 243–247.

Tagami, K., S. Uchida, N. Ishii, and J. Zheng, "Estimation of Te-132 Distribution in Fukushima Prefecture at the Early Stage of the Fukushima Daiichi Nuclear Power Plant Reactor Failures," *Environmental Science and Technology,* Vol. 47, No. 10, 2013, pp. 5007–5012.

Tagami, K., S. Uchida, Y. Uchihori, N. Ishii, H. Kitamura, and Y. Shirakawa, "Specific Activity and Activity Ratios of Radionuclides in Soil Collected About 20km from the Fukushima Daiichi Nuclear Power Plant: Radionuclide Release to the South and Southwest," *Science of the Total Environment,* Vol. 409, No. 22, 2011, pp. 4885–4888.

Takada, M., and T. Suzuki, "Early in situ Measurement of Radioactive Fallout in Fukushima City due to Fukushima Daiichi Nuclear Accident," *Radiat Prot Dosimetry,* Vol. 155, No. 2, 2013, pp. 181–196.

Takamura, N., and S. Yamashita, "Lessons from Chernobyl," *Fukushima Journal of Medical Science,* Vol. 57, No. 2, 2011, pp. 81–85.

Takemura, T., H. Nakamura, and T. Nakajima, "Tracing Airborne Particles After Japan's Nuclear Plant Explosion," *Eos,* Vol. 92, No. 45, 2011, pp. 397–398.

Tanabe, F., "Analysis of Core Melt Accident in Fukushima Daiichi-Unit 1 Nuclear Reactor," *Journal of Nuclear Science and Technology,* Vol. 48, No. 8, 2011, pp. 1135–1139.

Tanabe, F., "Analyses of Core Melt and Re-Melt in the Fukushima Daiichi Nuclear Reactors," *Journal of Nuclear Science and Technology,* Vol. 49, No. 1, 2012, pp. 18–36.

Tanabe, F., "A Scenario of Large Amount of Radioactive Materials Discharge to the Air from the Unit 2 Reactor in the Fukushima Daiichi NPP Accident," *Journal of Nuclear Science and Technology*, Vol. 49, No. 4, 2012, pp. 360–365.

Tanaka, Kazuya, Hokuto Iwatani, Aya Sakaguchi, Yoshio Takahashi, and Yuichi Onda, "Local Distribution of Radioactivity in Tree Leaves Contaminated by Fallout of the Radionuclides Emitted from the Fukushima Daiichi Nuclear Power Plant," *Journal of Radioanalytical and Nuclear Chemistry*, Vol. 295, No. 3, 2013, pp. 2007–2014.

Tanaka, Kazuya, Aya Sakaguchi, Yutaka Kanai, Haruo Tsuruta, Atsushi Shinohara, and Yoshio Takahashi, "Heterogeneous Distribution of Radiocesium in Aerosols, Soil and Particulate Matters Emitted by the Fukushima Daiichi Nuclear Power Plant Accident: Retention of Micro-Scale Heterogeneity During the Migration of Radiocesium from the Air into Ground and River Systems," *Journal of Radioanalytical and Nuclear Chemistry*, Vol. 295, No. 3, 2013, pp. 1927–1937.

Tanaka, K., Y. Takahashi, A. Sakaguchi, M. Umeo, S. Hayakawa, H. Tanida, T. Saito, and Y. Kanai, "Vertical Profiles of Iodine-131 and Cesium-137 in Soils in Fukushima Prefecture Related to the Fukushima Daiichi Nuclear Power Station Accident," *Geochemical Journal*, Vol. 46, No. 1, 2012, pp. 73–76.

Tanimoto, T., N. Uchida, Y. Kodama, T. Teshima, and S. Taniguchi, "Safety of Workers at the Fukushima Daiichi Nuclear Power Plant," *The Lancet*, Vol. 377, No. 9776, 2011, pp. 1489–1490.

Ten Hoeve, J. E., and M. Z. Jacobson, "Worldwide Health Effects of the Fukushima Daiichi Nuclear Accident," *Energy and Environmental Science*, Vol. 5, No. 9, 2012, pp. 8743–8757.

Thielen, H., "The Fukushima Daiichi Nuclear Accident-An Overview," *Health Physics*, Vol. 103, No. 2, 2012, pp. 169–174.

Thorp, J., "Fukushima Sequence of Events & Seismic Attributes," *Nuclear Plant Journal*, Vol. 29, No. 3, 2011, pp. 36–37.

"Toshiba Shows New Robot for Nuclear Cleanup," *CBC News*, November 21, 2012. As of August 2, 2013:
http://www.cbc.ca/news/technology/
toshiba-shows-new-robot-for-nuclear-cleanup-1.1131222

"Transactions of the American Nuclear Society and Embedded Topical Meetings 1st ANS SMR 2011 Conference and Young Professionals Congress 2011," Washington, D.C., 2011.

Tsuiki, M., and T. Maeda, "Spatial Distribution of Radioactive Cesium Fallout on Grasslands from the Fukushima Daiichi Nuclear Power Plant in 2011," *Grassland Science*, Vol. 58, No. 3, September 2012, pp. 153–160.

Tsuiki, M., and T. Maeda, "Spatial Variability of Radioactive Cesium Fallout on Grasslands Estimated in Various Scales," *Grassland Science,* Vol. 58, No. 4, 2012, pp. 227–237.

Uchida, S., and Y. Katsumura, "Water Chemistry Technology - One of the Key Technologies for Safe and Reliable Nuclear Power Plant Operation," *Journal of Nuclear Science and Technology,* Vol. 50, No. 4, 2013, pp. 346–362.

U.S. Environmental Protection Agency, Office of Air and Radiation, *Technology Reference Guide for Radiologically Contaminated Surfaces*, Washington, D.C., EPA-402-R-06-003, April 2006.

Van Deventer, E., M. Del Rosario Perez, A. Tritscher, K. Fukushima, and Z. Carr, "WHO's Public Health Agenda in Response to the Fukushima Daiichi Nuclear Accident," *Journal of Radiological Protection,* Vol. 32, No. 1, 2012, pp. N119–N122.

Von Hippel, F. N., "The Radiological and Psychological Consequences of the Fukushima Daiichi Accident," *Bulletin of the Atomic Scientists,* Vol. 67, No. 5, 2011, pp. 27–36.

Vuddhakul, V., T. Nakai, C. Matsumoto, T. Oh, T. Nishino, C. H. Chen, M. Nishibuchi, and J. Okuda, "Analysis of gyrB and toxR Gene Sequences of Vibrio Hollisae and Development of gyrB- and toxR-Targeted PCR Methods for Isolation of V. Hollisae from the Environment and Its Identification," *Applied and Environmental Microbiology,* Vol. 66, No. 8, 2000, pp. 3506–3514.

Wang, Hui, ZhaoYi Wang, XueMing Zhu, DaKui Wang, and GuiMei Liu, "Numerical Study and Prediction of Nuclear Contaminant Transport from Fukushima Daiichi Nuclear Power Plant in the North Pacific Ocean," *Chinese Science Bulletin,* Vol. 57, No. 26, 2012, pp. 3518–3524.

Westlake, Adam, "Robot Suit to Be Used in Fukushima Cleanup, Controlled by Brain Waves," *Japan Daily Press,* October 18, 2012. As of August 2, 2013: http://japandailypress.com/robot-suit-to-be-used-in-fukushima-cleanup-controlled-by-brain-waves-1816592/

Woo, T. H., "Atmospheric Modeling of Radioactive Material Dispersion and Health Risk in Fukushima Daiichi Nuclear Power Plants Accident," *Annals of Nuclear Energy,* Vol. 53, 2012, pp. 197–201.

Woods, V. T., T. W. Bowyer, S. Biegalski, L. R. Greenwood, D. A. Haas, J. C. Hayes, E. A. Lepel, H. S. Miley, and S. J. Morris, "Parallel Radioisotope Collection and Analysis in Response to the Fukushima Release," *Journal of Radioanalytical and Nuclear Chemistry,* Vol. 296, No. 2, 2013, pp. 883–888.

Yamaguchi, K., and University Radiation Survey Team of Fukushima, "Investigations on Radioactive Substances Released from the Fukushima Daiichi Nuclear Power Plant," *Fukushima Journal of Medical Science,* Vol. 57, No. 2, 2011, pp. 75–80.

Yamaguchi, N., S. Eguchi, H. Fujiwara, K. Hayashi, and H. Tsukada, "Radiocesium and Radioiodine in Soil Particles Agitated by Agricultural Practices: Field Observation After the Fukushima Nuclear Accident," *Science of the Total Environment,* Vol. 425, 2012, pp. 128–134.

Yamaguchi, T., K. Sawano, M. Kishimoto, K. Furuhama, and K. Yamada, "Early-Stage Bioassay for Monitoring Radioactive Contamination in Living Livestock," *Journal of Veterinary Medical Science,* Vol. 74, No. 12, 2012, pp. 1675–1676.

Yamamoto, S., and J. Hatazawa, "Development of an Alpha/Beta/Gamma Detector for Radiation Monitoring," *Review of Scientific Instruments,* Vol. 82, No. 11, 2011.

Yasunari, T. J., A. Stohl, R. S. Hayano, J. F. Burkhart, S. Eckhardt, and T. Yasunari, "Cesium-137 Deposition and Contamination of Japanese Soils due to the Fukushima Nuclear Accident," *Proceedings of the National Academy of Sciences of the United States of America,* Vol. 108, No. 49, December 2011, pp. 19530–19534.

Yemelyanau, Maksim, Aliaksandr Amialchuk, and MirM Ali, "Evidence from the Chernobyl Nuclear Accident: The Effect on Health, Education, and Labor Market Outcomes in Belarus," *Journal of Labor Research,* Vol. 33, No. 1, 2012, pp. 1–20.

Yoshida, K., K. Hashiguchi, Y. Taira, N. Matsuda, S. Yamashita, and N. Takamura, "Importance of Personal Dose Equivalent Evaluation in Fukushima in Overcoming Social Panic," *Radiation Protection Dosimetry,* Vol. 151, No. 1, 2012, pp. 144–146.

Yoshida, N., and J. Kanda, "Tracking the Fukushima Radionuclides," *Science,* Vol. 336, No. 6085, 2012, pp. 1115–1116.

Yoshihara, Toshihiro, Hideyuki Matsumura, Shin-nosuke Hashida, and Toru Nagaoka, "Radiocesium Contaminations of 20 Wood Species and the Corresponding Gamma-Ray Dose Rates Around the Canopies at 5 Months After the Fukushima Nuclear Power Plant Accident," *Journal of Environmental Radioactivity,* Vol. 115, 2013, pp. 60–68.

Zheng, J., K. Tagami, Y. Watanabe, S. Uchida, T. Aono, N. Ishii, S. Yoshida, Y. Kubota, S. Fuma, and S. Ihara, "Isotopic Evidence of Plutonium Release into the Environment from the Fukushima DNPP Accident," *Scientific Reports,* Vol. 2, March 2012.